HOW THE BRAIN PROCESSES MULTIMODAL TECHNICAL INSTRUCTIONS

Dirk Remley
Kent State University

Baywood's Technical Communications Series
Series Editor: CHARLES H. SIDES

Baywood Publishing Company, Inc.
AMITYVILLE, NEW YORK

Printed in the United States of America on acid-free recycled paper.

Baywood Publishing Company, Inc.

26 Austin Avenue
P.O. Box 337
Amityville, NY 11701
(800) 638-7819

E-mail: baywood@baywood.com
Web site: baywood.com

Library of Congress Catalog Number:
ISBN: 978-0-89503-893-7 (cloth : alk. paper)
ISBN: 978-0-89503-894-4 (paper)
ISBN: 978-0-89503-895-1 (e-pub)
ISBN: 978-0-89503-876-8 (e-pdf)
http:/dx.doi.org/10.2190/HTB

Library of Congress Cataloging-in-Publication Data

Remley, Dirk.
How the brain processes multimodal technical instructions / Dirk Remley, Kent State University.
pages cm .- - (Baywood's technical communications series)
Includes bibliographical references and index.
ISBN 978-0-89503-893-7 (clothbound : alk. paper) -- ISBN 978-0-89503-894-4 (paperbound : alk. paper) -- ISBN 978-0-89503-895-1 (e-pub) -- ISBN 978-0-89503-876-8 (e-pdf)
1. Technical communication. 2. Multimodal user interfaces (Computer systems) 3. Reasoning. I. Title.
T11.R42453 2015
601'.9—dc23
2015026961

Table of Contents

Preface

The fields of cognitive psychology and neuroscience integrate several disciplines ranging from natural sciences such as biology, and physical sciences such as physics, to social sciences such as psychology and rhetoric, which may also be considered within the humanities. Neuroscience has found that learning and cognition are multisensory experiences, that is, no one sense is involved in learning something new. Whenever one tries to theorize a concept that may be interdisciplinary in nature, such as learning, the temptation is to consider the theories closest to the researchers' own discipline while omitting consideration of those outside of the discipline. This is, at times, both necessary and limiting.

It may be necessary to focus on a single theoretical perspective merely to help manage the research method and analysis. One skilled with social research methods such as surveys or quasi-experimental designs cannot use tools that measure biological functions such as neural activity as well as one trained to use tools such as fMRI machines. Likewise, one whose field values fMRI studies will not pursue social science research knowing that it may not be published by journals esteemed in their own discipline.

However, such disciplinary focus also limits the scope with which the phenomena associated with the subject of study can be studied. A social scientist is not able to account for biological attributes involved in a particular phenomenon. Likewise, a natural scientist may not be able to account for social attributes involved in the particular phenomenon. These limit our understanding of the phenomena.

Many scholars are calling for interdisciplinary research into cognition. However, disciplinary discourses, which include value systems, interfere with such efforts. This book represents an effort to identify a means to engage multiple fields associated with the cognitive function of learning. I try to integrate attributes of theories from these different fields into a single model, recognizing that these attributes affect learning and that ignoring any of them limits the study of learning.

I am most familiar with the fields of multimodal rhetoric and semiotics. Rhetoric generally recognizes that biological attributes affect one's understanding of a message. However, I am not trained as a biologist. In an effort to engage biological attributes of rhetoric, I have studied cognitive neuroscience literature, recognizing my own limitations in fully understanding it due to very specialized language. In spite of discourse differences, this research reinforces many findings reported in studies of multimodal rhetoric. Further, it helps to explain the biological attributes related to many of those findings.

As the series in which this book is published suggests, this book targets a particular audience—technical communicators, many of whom are also involved with studies of multimodal rhetoric and semiotics. Consequently, I simplify some terminology from neurobiology while staying true to the disciplines; I have consulted with a neurobiologist to ascertain that I have used terms accurately and represented that field's work reasonably. However, I recognize that others outside of these fields may read it as well. I welcome and caution those other readers.

I do not integrate scholarship from some of these fields, recognizing my own limitations. I also recognize this model as an early effort to encourage and facilitate the kind of interdisciplinary research for which so many have called. It is a starting point, not an end. The model proposed and explained here is descriptive rather than prescriptive. As such, it invites further multidisciplinary development. I encourage readers from any field to consider the model as a foundation for analysis in their own interdisciplinary research and add to it with findings from their work. Much as the periodic table of elements has grown with discovery or development of new elements, this model will grow with the integration of disciplinary perspectives outside of those considered here and toward a more comprehensive interdisciplinary model.

It is my hope that this model is generative for interdisciplinary work in the study of the cognition of learning. I know that scholarship in multimodal rhetoric has informed improvements in the design of instructional materials as new technologies have emerged. I also observe that scholarship in neuroscience and cognitive psychology have further informed this improvement in instructional materials. However, a model that bridges gaps between fields while recognizing the value of each field is needed. This book represents an effort to create such a model.

Acknowledgment

I would like to thank Dr. Jeffrey Wenstrup, Professor and Chair of Anatomy and Neurobiology at the Northeast Ohio Medical University, for his insight and review of a few of the chapters in this book to ascertain that I used neurobiological concepts accurately. While reading several articles and book chapters on neuroscience/neurobiology related to multimodal stimuli helped me to ascertain connections between the fields for the theorization I develop in the book, it is through review of application of those concepts and information that one demonstrates their understanding of the material. Further, as those in writing studies already understand, it is through feedback that one learns how well they understand and can improve their understanding of material.

CHAPTER 1

A Gap to Bridge

How does one learn a new technical concept? A teacher or trainer can provide information about that new concept to a student or new employee; he or she can even show the student or employee how to practice or apply the concept. However, what happens in the student's mind to help him or her learn the concept and its application? Learning involves many attributes and dynamics generally associated with cognitive psychology, education, semiotics, rhetoric, and neurobiology. That is, the ways one presents instructional information and the student's natural biological attributes associated with processing that information affect how the student learns. However, these are disparate fields of scholarship, and they pursue study of the question, "How does one learn a new technical concept?" differently.

Neurobiologists Calvert, Spence, and Stein (2004) noted that, because the scholarship related to neuroscience is "spread across multiple disciplines, it has become increasingly fragmented in recent years" (p. xii). However, in a special issue of *Technical Communication Quarterly*, Rivers (2011) encouraged a multidisciplinary approach to research into cognitive science, recognizing the roles that biology and social environment as well as technology play in cognition. Alluding to the convergence of tools, environment, and brain in distributed cognition, he stated that, "those tools and that world are always part of the mind itself" (p. 415). With this book, I attempt to synthesize some scholarship in disparate fields to provide a bridge by which interdisciplinary study that may enhance the development of instructional materials may occur. I introduce a model by which to study the question, "How does one learn new technical concepts?" This model recognizes that learning involves social, biological, and rhetorical attributes that engage many senses. Consider these examples:

It is a frequent experience—having to drive to a doctor's office for the first time. You contact the doctor's office, and the receptionist gives you directions to the office, usually identifying particular landmarks—a restaurant or shopping plaza near it and approximate distances from those items to the office building—"about 100 yards after that restaurant, you'll cross an intersection. On the right you'll see a gas station; our building is right across the street from that"—perhaps even the color of the building—"it's grey." You get into your car on the big day of the appointment, and you pay close attention to these visual and spatial cues, looking carefully for the restaurant, seeing it, and understanding that you are very close to the building. You suddenly see a gas station and you look across the street; there it is—the grey building. Using the information—visual landmarks and spatial relationships—you succeeded in finding the building. Now you have only to navigate the building to find the doctor's office!

Another frequent experience is that of hearing the blare of sirens from an approaching emergency vehicle before you see it. Sometimes one may never actually see the vehicle—fire engine, ambulance, or police car—understanding when it passes the vicinity, though, from the changing sound of the blaring siren. Sometimes one sees the vehicle a few moments after hearing the sirens and moves their vehicle aside or watches it pass. In either case, one has learned what such a signal generally means and understands that there is an emergency somewhere and acts accordingly. Whether one sees it or not, an audio message is sent, and a listener reacts to it before seeing and may react again upon seeing it. I experienced some surprise when I heard and then saw a fire engine approach my house, though I understood partially the reason. The implication, though, raised further questions that prompted more multisensory experiences connected to an emergency. I relate the experience later in this chapter.

In each of these cases, though, different senses were engaged to help one understand a situation. A number of dynamics related to one's understanding of the world, cognition, and forms of representation played out so that person could understand and act accordingly. There is a multimodal rhetoric to these experiences. The field of "rhetoric" generally engages the question of how one may convey a message, either in writing or orally, to bring about a particular response from an audience. The field of multimodal rhetoric recognizes that one may convey their message using multiple forms of representation, thereby engaging multiple senses in an audience and creating meaning with combinations of these modes of representation. However, which combinations most effectively facilitate learning depends on a number of social, biological, and cognitive attributes associated with the learner.

The field of social semiotics, further, recognizes that meaning is a social construct, that is, one's interpretation of various images and objects evolves through interactions with others. One has certain sensory information provided to them based on experiences and knowledge that likely involve social settings and interactions with others: You understand the message and you respond to the message according to its purpose.

The cognitive experience is rhetorical and social. We learn about new concepts by being exposed to information about them and interacting with phenomena associated with the new concepts. This interaction may be through discussion or practical experience. One may provide information to us in a way that will help us to understand a new concept; this is both a social phenomenon—interaction with another—and it is rhetorical—a message is provided to an audience (us) with the purpose of helping us understand something new. It takes some interaction with the world around us to comprehend a situation and the meaning of the information provided. However, it is also biological. Cognitive science generally recognizes these attributes of cognition—social and biological attributes related to facilitating an understanding of our world. However, the discussion of these cognitive neuroscience dynamics is complicated by disciplinary discourses and exclusions.

Each discipline approaches the topic from its own angle, recognizing that literature from that field is needed to support such scholarship. For example, rarely will the author of a scholarly article cite work from outside their own discipline or the discipline of the particular journal. This extends to scholarly books too. For example, in his highly regarded book *Cognition in the Wild*, Hutchins (1995) limited the discussion of cognition and social semiotics to cognitive psychology and distributed knowledge theory. Also, in *How the Mind Works*, another highly regarded work of cognitive neuroscience, Pinker (1997) integrated some discussion of neuroscience on cognitive processes, however, he focused on historical development of cognitive processes and psychological evolution. Finally, Gruber (2012) highlights discourse differences in how scholars treat the neuroscientific concept of mirror neurons, neurons that help an audience interpret and copy behavior they view. Such discourse exclusion limits the lens through which studies examine the phenomena.

Further, approaches to empirical study affect the lens through which scholars view phenomena. Typically, empirical study of learning from social science perspectives involves presenting subjects with instructional materials, having them review those materials, and then assessing their performance on a test or other means of learning assessment to ascertain how much they learned from the instructional materials (see studies by Moreno & Mayer, 2000, for examples). Studies from physical science perspectives may ask subjects to review instructional materials while the subject wears equipment that measures and facilitates observation of electrical or blood activity in the brain to ascertain which parts of the brain are involved in processing information (see studies reported in Calvert et al., 2004, for examples). While social scientists appear to be studying how well certain instructional materials facilitate learning—what works and what does not—physical scientists are examining the biological dynamics of cognition and how certain biological attributes may affect learning—why certain approaches work and others do not.

The purpose of this book is to address some of these differences so that scholars in both fields can pursue interdisciplinary research that may enhance our understanding of cognitive psychology and neuroscience generally. Study that recognizes the value of considering what works as well as why it works is important toward facilitating a more comprehensive consideration of cognition.

Specifically, I describe a model that integrates neurobiological attributes involved in cognition and social attributes involved in learning new information provided using multiple modes of representation. I refer to it as a neurocognitive model of multimodal rhetoric or as an integrated model of cognition. I focus on it specifically relative to instructional materials and how they can help an audience learn. Attributes of cognitive psychology and neuroscience engage many different functions, including decision making as well as changing perspectives and management (Pillay, 2011). Some of these overlap with learning, so I integrate them here. Any form of communication and rhetoric includes consideration of the purpose of the message. Instruction is just one of many purposes. The focus of the model is on how the brain processes new technical information or concepts toward learning them. The model recognizes that some with a given neurological condition may need information to be provided in ways that differ from how it is presented to others without that condition, but the information and how to apply it can be understood and learned by both groups.

Rhetoric, in a broad sense, examines how the way information is presented affects an audience's understanding of that information and response to it. Aristotle (1991) and Perelman and Olbrechts-Tyteca (1969) recognized that rhetoric considers the disposition of certain kinds of audiences, and one who wishes to convey an effective message must adjust to their particular audience. Aristotle acknowledged that rhetoric includes "three factors—the speaker, the subject and the listener—and it is to the last of these that its purpose it intended" (p. 80). The purpose of a message and its audience are intertwined. The message must consider the audience's disposition in order to accomplish its purpose. This disposition can be theorized relative to social disposition or biological/physical disposition. Indeed, Aristotle noted that this likely involves an audience that may have "limited intellectual scope and limited capacity to follow an extended chain of reasoning" (p. 76). Such a statement includes physiological attributes in the rhetoric equation. If the audience's cognitive capacities are not considered in developing the message, the meaning of the message will be lost.

Perelman and Olbrechts-Tyteca (1969) recalled this emphasis on the audience, pointing out that "it is in terms of an audience that an argumentation develops" (p. 5). Indeed, they compared one who does not consider the audience to a rude visitor (p. 17). They asserted that the most important rule of rhetoric is to adapt the message to the audience (p. 25). A message is not automatically understood just because it is articulated; it must be conveyed in a way that suits the audience's background and understandings, their experiences and practices, their capacity for cognition.

Scholarship in rhetoric draws on studies from the social science and humanities disciplines of cognitive neuroscience—social semiotics, social psychology, and language theories. Rhetoric is certainly a social dynamic. However, rhetoric has been left out of much of the discussion of cognitive neuroscience and is not considered among those fields. Jack (2012) as well as Jack and Appelbaum

(2010) provide some introductory material for connecting rhetoric with biological fields of neuroscience in her edited collection about "neurorhetorics." Jack and Appelbaum (2010) identified two approaches to "neurorhetoric." One involves studying the rhetoric of neuroscience, in which one considers how different discourses treat neuroscientific scholarship. Gruber (2012), for example, takes the first approach and describes the discourse differences related to how different fields treat a particular concept of neuroscience—the concept of "mirror neurons." He observes that institutional dynamics at work within disciplinary scholarship limit the ability to arrive at a common language to describe the concept, further illustrating this problem. They also state that a

> second approach might be the neuroscience of rhetoric, drawing new insights into language, persuasion, and communication from neuroscience research. Findings such as this study of noncommunicative patients can prompt us to broaden our very definitions of rhetoric to include those with impaired communication (such as autism, aphasia, or "locked-in syndrome"), asking how communication occurs through different means, or how brain differences might influence communication. (Jack & Appelbaum, 2010, p. 412)

I attempt to close some of the discourse disconnections Gruber (2012) and Calvert et al. (2004) identified while using the second approach to synthesize scholarship in multimodal rhetoric and neurobiology, particularly with respect to multisensory neural processes, explicitly in the discussion of cognitive neuroscience.

Gruber (2012) formulates four "pillars" by which interdisciplinary research involving rhetoric and neuroscience can occur by facilitating a means of "translation" between discourses. These pillars are very much a building tool applied in this book. The first pillar, he explains, is the "field-familiar spokesperson" (p. 237). This is a scholar who is knowledgeable about neuroscience and a second field—a sort of intermediary between discourses. I represent this person in the context of this book. The second pillar is that of the spokesperson's support—a mechanism by which the spokesperson from the first pillar establishes ethos, or credibility, as well as logos for establishing the connection with the neuroscience community (p. 239). As I explained in the Author's Preface, I have consulted with a neurobiologist to ascertain that I understood concepts of neurobiology that I present in this book and applied them correctly. This neurobiologist acts as the second pillar in the context of this book.

The third pillar is that of nature; Gruber (2012) indicates that nature connects neuroscience with the particular field being applied to it or vice versa. I have already alluded to Aristotle's (1991) and Perleman and Olbrechts-Tyteca's (1969) references to the links between rhetoric and biology. These and the scholarship in neurobiology that I cite contribute to establishing this pillar for this book. The last pillar Gruber identifies is that of "objective writing practice."

He explains that this is a practice that makes writing transparent rather than an exercise in creativity; it is an effort to represent an objective reality rather than corrupt reality.

The model that I propose and develop here rests on these pillars and is open to further construction. As scholars in rhetoric and other disciplines interact with this model, they act as additional field-familiar spokespersons, lending their credibility to the model's development and applications. When two or more researchers from different fields join to study a given phenomenon, a synergistic effect occurs within the dynamic of those pillars to strengthen the model and allow for further development.

Cognitive Neuroscience and Rhetoric

The field of neuroscience has experienced a boom in scholarship that integrates several disciplines. Generally, this scholarship ranges across the five general disciplines that are connected with cognitive neuroscience: cognitive psychology, philosophy, linguistics, biology, and chemistry. Physics is also somewhat involved. Most of these are recognized as humanities-related areas, while the others are specifically connected to science—biology, chemistry, and physics. As mentioned above, each discipline theorizes neuroscience and cognition by applying its own research methods and theories to analysis and discussion. However, the disconnection across disciplines is problematic, especially as institutions attempt to find ways to connect disciplines with interdisciplinary programs and research projects. Cognition is associated closely with perception: How one perceives information affects their understanding of that information. The field of cognitive neuroscience devotes much attention to understanding how one processes information toward cognition. Humanities scholars tend to examine how language and social interactions affect our understanding of the world. Reid (2007) noted that "cognitive scientists termed the 1990s 'the Decade of the Brain' for the startling advances made throughout their discipline" (p. 14). Indeed, Hutchins (1995) and Pinker (1997) theorized cognition as a series of developmental processes that include historical dynamics as well as how people treat training and actual practice and the social dynamics therein. This has helped to generate subfields of distributed cognition and cognitive psychology as well as social semiotics. In each case, research in cognitive neuroscience has found that cognition is a multisensory process. Social interaction engages multiple senses—visual, aural, spatial orientation, and relationship—as well as gesture—touch and smell. Likewise, language is generally recognized as being aural/oral or visual, and print-linguistic text is a visual representation.

Science disciplines have been studying connections between perception, behavior, and neural dynamics. Available technology affects how this study occurs. Until recently, most of this involved looking at electrical activity within the brain. Neurons send electrical messages across the brain, and the different parts process that information toward doing something with it. However, recent technology has made it possible to look into other physical attributes of the brain and how the brain processes information related to perception and cognition. In particular, 2-photon microscopes and magnetic resonance imaging (MRI) technology facilitates such research. A 2-photon microscope permits the imaging of areas of the brain that are excited during tasks, suggesting neural activity. Some MRI technology allows researchers to see how blood flows to certain parts of the brain while one performs a particular task—viewing a given film or doing certain work, for example. This technology is called "functional MRI," or "fMRI." Biologists and chemists have begun examining the relationship between blood flow and neural processes. As humanities scholarship has done, many of these studies also link cognition to multisensory processes (e.g., see collections edited by Calvert et al., 2004, and Murray & Wallace, 2012).

Rhetoric encompasses a range of communication practices, including informational messages, persuasive messages, and instructional messages. However, my focus in the book is on the neurorhetoric of instruction and learning. Some studies have found that persuasion involves some different neural activities than cognition related to cognition does (Azar, 2010; Pillay, 2012; Ramsay et al., 2013). There is more self-reflection and reflection about one's perception of others and attitudes. Persuasion is a belief-oriented or attitude-oriented concept. The general focus of persuasion is to change one's attitude or beliefs about a given topic or issue or to elicit a stronger conviction in belief or attitude about that topic or issue. While mirror neurons, for example, are involved in this process as well, that involvement has more to do with mirroring or sharing a perception ("shared emotion") than with copying or imitating action. Pillay (2012) points out that "our brains can mirror not only actions, but intentions as well" (p. 63). When a manager or supervisor seems to treat a situation as negative, subordinates seem to perceive it similarly, as their brain mirrors the supervisor's perception of the situation. The field of rhetoric has begun to study this empathetic activity, however, it is different from the kind of rhetoric associated with the scope of this book.

The field of multimodal rhetoric, especially, considers how various modes of representation affect an audience's ability to make meaning about the information provided with those combinations. Rhetoric is already linked implicitly to the social sciences disciplines of cognitive neuroscience through existing scholarship related to social semiotics and cognition (e.g., Gee, 2003; Hutchins, 1995; Kress, 2003; Moreno & Mayer, 2000; New London Group, 1996; Tufte, 2003). I connect scholarship of multimodal rhetoric explicitly to the scientific fields of cognitive neuroscience relative to cognition associated with learning. Also, I argue that scholarship in rhetoric can inform studies related to other areas of cognitive neuroscience and learning, contributing to development of productive instructional materials as well as to biomedical advances in cognition and brain development. The fields are already somewhat connected by terminology, which can act as a bridge to facilitate interdisciplinary research.

Neurobiologists refer to neural processes that integrate multiple senses as "multisensory integration." This term is the biological equivalent of the term "multimodal" in rhetoric. In fact, neurobiologists sometimes refer to multisensory integration as multimodal integration. While multimodal in rhetoric pertains to multiple modes of representation used to convey a message, multisensory integration refers to the interaction of multiple senses used at the same time to process information. In neuroscience, the modes of representation are treated as stimuli from which neurons initiate information processing. So a visual representation—a photograph, for example—would stimulate neurons associated with visual information processing. Likewise, a message that included visual and audio forms of representation, like a video with narration, would be considered a form of multimodal rhetoric from a rhetoric perspective, and it involves multimodal integration from a neurobiological perspective. This link is natural and logical for two other reasons.

Rhetoric and Science

Rhetoric is already linked and theorized relative to the humanities and social sciences. While this link will continue to be pursued (and should be pursued), it is important to recognize biological attributes that affect rhetoric. As Reid (2007) observed, "All information, even speech, enters our body in analog form" through various sensory experiences (p. 17). Neurons help to process it toward facilitating meaning-making and cognition of the information. However, rarely are biological attributes of cognition factored into the discussion of rhetoric. Scholarship in rhetoric tends to focus on the product and its effect, not the process that affects that effect. Studies of brain disorders and how such disorders or injuries affect learning are occurring. These studies benefit education and training programs by facilitating an understanding of how to design instructional materials so that people with certain cognitive limitations may be able to learn concepts and tasks. An understanding of the process can contribute to designing more effective messages.

So one reason to encourage a closer connection between rhetoric and science is to recognize that biological dynamic that affects how an audience responds to a message. Many studies of multimodal instructional design and learning within the fields of educational psychology and multimodal rhetoric, such as those by Moreno and Mayer (2000), consider how students demonstrate their learning with particular instructional materials by examining performance. That is, they provide a starting point (instructional materials, or the message) and measure the end point (performance, or what the audience learned from the message). Leaving biology out of rhetoric is like a writing teacher grading an essay based only on the product and not caring about the writing process. Any writing teacher recognizes that practice as bad pedagogy. It is generally recognized in writing studies that if a writing teacher understands the process by which the student composed the essay, he or she can provide much better instruction to help improve the student's writing process toward improving their writing than by focusing only on a writing product and identifying errors.

People learn to do things through experience with a task. The brain facilitates learning through neural processes and grows neural features with each experience. Indeed, scholarship in neurobiology studies the changes the brain experiences as it learns new tasks and information. I argue that rhetoric scholarship should integrate more consideration of biological attributes and dynamics that affect an audience's perception of a message. Further, I argue that a synthesis of the discourses to facilitate interdisciplinary scholarship is possible.

A second reason for my theorization relative to science is political in nature, recognizing the current academic environment. Academic programs associated with science, technology, engineering, and math (STEM disciplines) are receiving higher consideration for funding and marketing. Social science and humanities disciplines have to defend their relevance in the current environment, and many are facing reduced funding. It is important to show the relevance of rhetoric to this STEM "fever" so that interdisciplinary research that recognizes the value of rhetoric can occur. However, this relevance is not superficial; it is very much credible and valid. Rhetoric can make an impact in science-oriented programs. A text making this connection, introducing a discourse that integrates language from both fields and talking about contributions of such research and their impact can introduce such interdisciplinary scholarship.

I theorize multimodal rhetoric pertaining to learning through weaving neuroscientific theory, particularly referring to neurobiological studies associated with electrophysiology (patterns of electrical flow within the brain) and hemodynamic attributes of brain processes (studies of blood flow to certain parts of the brain associated with sensory experiences and cognition) with multimodal theory. Using existing theories of multimodal rhetoric and studies related to electrophysiology and the hemo-neural hypothesis, I develop a discourse model that integrates elements from both fields uniting them. This theorization is easily facilitated because much of what scholarship in multimodal rhetoric has found is reinforced in scholarship connected to neural processes. Indeed, neural process studies help to explain some of the biological attributes connected to findings of multimodal scholarship, that is, it helps us understand *why*, from a biological perspective, certain multimodal products facilitate a better understanding of information than other multimodal products.

Through synthesizing multimodal rhetorical theory with what is understood about how the mind processes information related to learning tasks and concepts relative to its use of blood, it is possible to refine that theory as it pertains specifically to training/instruction and process improvement, two topics of considerable interest in both industry and education. I review extant theories of multimodality and neuroscience—particularly related to two methods associated with biophysiological analysis—electrophysiology and the hemo-neural hypothesis. I also discuss workplace and educational applications of the emerging model, particularly cases that integrate approaches popular in education and training such as experiential learning and new media. I devote one full chapter to a particular model of training esteemed in the lean operating philosophy currently valued in business and industry. The Training Within Industry (TWI) program integrates all modes of representation as characterized by the New London Group (1996). These include printed, alpha-numeric text, oral/aural, visual, behavior/experiential, and combinations of these, much as experiential learning and service learning do. Because all are valued in the TWI model, it serves as a vehicle by which readers can better understand the links between multimodal rhetoric and cognitive neuroscience. I also discuss cases related to new media applications found in education and training, including simulators, instructional slide shows, and instructional video. All of these integrate various combinations of modes of representation and media, engaging multiple senses.

With each of these applications, I discuss the interrelationships of the various modes of representation associated with cognition. I build on existing multimodal learning theory (especially that of Moreno & Mayer, 2000, and Tufte, 2006), and I introduce new concepts of visual rhetoric relative to scholarship in cognitive neuroscience and how that scholarship theorizes interconnections of multisensory stimuli. While the previous examples I mentioned to illustrate some attributes of cognition were brief, I elaborate on the one involving the fire engine to further illustrate my point about cognition being a multisensory experience.

A MULTIMODAL COGNITIVE EXPERIENCE INVOLVING AN EMERGENCY

The fire engine experience I mentioned earlier involves two particular events: the trip from the fire engine and a call with a retailer. It features a range of multimodal rhetorical messages that integrate smell, audio, visual, and other sensory experiences. It also includes social as well as biological attributes of cognition.

Some years ago, I purchased a gas fireplace for my house. My house was heavily dependent on electric heat, which was okay, but I felt the need to have gas heat available if the power ever went out for an extended period during the winter months. Such a fireplace also facilitated ambiance on a nice quiet evening. So I felt very good about the purchase, and I looked forward to its installation. It was installed a few afternoons later, and the installers acknowledged that there would be a somewhat unpleasant smell as some materials burned off during the initial operation of the fireplace. Sure enough, there was a certain odor in the first hours of burning. After the initial full day of burning, though, I understood that the odor would be gone, yet it still lingered a bit. I also had other unpleasant senses the second day—dryness in my mouth and some lightheadedness.

The Fire Engine

The second morning after installation (a Sunday), particularly concerned about the dry mouth and lightheadedness, I dialed the nonemergency phone number of the local fire department and acknowledged the installation of the fireplace and my sensory experiences. They reassured me that they would send someone over to take a look. A few minutes after I hung up, I heard the siren that one generally associates with an emergency vehicle, figuring it was making its way to an emergency somewhere in the city. I noted it, and went about my business waiting for someone to arrive to look at the fireplace; I had called the nonemergency number, so it *clearly* was not related to me.

As I waited, I noted that the siren was getting closer to me, and I figured that the emergency was nearby or the vehicle had to pass near my location to get to the site. However, it got closer and closer to me. With each second, it got closer. I first figured that someone in the same development as mine had an emergency; this was new and interesting. It got closer. I figured one of my neighbors must have an emergency—I grew concerned for my neighbors.

Then, with the siren getting even louder to the point one could not hear another talking, I *saw* a fire engine turn the corner onto my street, and I awkwardly hoped one of my neighbors had an emergency. I had called the nonemergency number, but here was an emergency vehicle making its way toward my house. Sure enough, it stopped right at my house.

I went outside to greet the firemen with the question, "What are you doing here?! I called the nonemergency number!" The first fireman to get out of the vehicle acknowledged, "Well, we're not sure what we might find once we get here." My mind reeled! So my house and I might have been blown up while waiting?

As the fireman entered the house, I told him some details of the burning and the sensory experiences I had. He looked at the fireplace and immediately stated that the flame should be bluer than it was. The flame had very aesthetic-looking yellow and orange colors in it, much like a "real" fire would, as the marketing materials for the fireplace acknowledged. However, as a gas fireplace, this was a problem—it suggested a poor mixture of oxygen in the burning process, which could lead to carbon monoxide poisoning. I saw the very same visual information he had, yet he was able to identify a problem with the fireplace that explained my symptoms that I could not have identified. He had background with and knowledge about flame colors, gas products, and chemical mixtures that I did not have. This information helped him to identify a particular problem that explained my own symptoms of carbon monoxide poisoning. He encouraged me to get a carbon monoxide (CO) detector and let the retailer know about the problem.

I bought the detector and plugged it in about 10 feet from the fireplace. According to the information on the manual (yes, I actually read the manual) it would generate a number every 3 minutes indicating the carbon monoxide content per cubic meter, and if it went above a certain number (50 ppm), there was too-high a level of CO. Further, if it went above a particular number above that which represented an emergency CO situation (100 ppm), it would sound an alarm. Within an hour of plugging it in, it reached the level indicating too-high level, and I opened the windows of my house and noted to contact the retailer the next morning (Monday).

The Retailer

The next day, I ran the fireplace again and had the CO detector operating to monitor CO levels. As the readout rose above the too-high level, I let the fireplace continue to burn to see how high the reading might go. As it approached the emergency situation level, I phoned the fireplace retailer about my experience over the weekend, acknowledging the visit from the fire department and the CO detector results. I also indicated the status of the current "test."

The retailer did not convey much concern, indicating that CO detectors tend to be sensitive and that CO levels indicated are not as dangerous as indicated on the manual, mitigating the relevance of the readout. As I was on the phone with him, though, I noted that the number was very close to the number at which the CO detector alarm would sound, but the retailer continued to minimize the validity of the detector. However, when the alarm sounded, the retailer could clearly hear it. He said he would be right over to look at the fireplace and check out the problem, and he was in my house within 15 minutes of the alarm sounding.

MULTISENSORY RHETORIC AND NEUROSCIENCE

I present these anecdotes because they illustrate the relationship between various sensory experiences and rhetoric thereof. While instruction was not a primary goal of each experience, there is an attribute of learning associated with them. Further, biological attributes as well as social semiotic and distributed cognition dynamics come into play in each to help people understand a given message. In both cases, visual information was used to refine an understanding of auditory information. Further, my own experience with smell, taste, and lightheadedness caused me to be concerned about the situation and contact someone who would know more than I did. It provided me information that something was wrong and needed to be addressed by someone who knew more than I did about particular dynamics of gas fireplaces.

In both cases, a professional came to my house to see the fireplace and experience visual information to help address a problem. The fireman used it to ascertain a problem with the burner to explain my biological symptoms, and the retailer used it to ascertain more specifically the problem so he could find repair materials to address it. Further, from my interaction with the fireman and additional research based on that interaction, I came to learn more about visual information provided by colors of flames in a gas fireplace set. Visual rhetoric has emerged as a field of study, and texts like Petroski's *Invention by Design* and Tufte's *Beautiful Evidence* (2006) detail applications of visual rhetoric. However, neither describes *why* visuals tend to be esteemed.

The retailer also was persuaded to look into the problem based on the audio information provided by the CO detector. He had been minimizing the value of the CO detector while I described the situation to him over the phone, yet once he heard the alarm sound, he was persuaded to attend to the problem.

Several articles detail various applications of multimodal rhetoric relative to a given theory and specific case. Works cited often in articles on multimodal literacy practices and rhetoric include those by Kress (2003), Kress and van Leeuwan (2001), as well as the New London Group (1996); these discussed rhetoric relative to social semiotic theory. Other works discuss applications of particular kinds of multimodality without showing interrelationships across the different types and modes.

Mayer (2001) discussed experimental studies of multimedia instruction and theorized about multimodality within such contexts. However, he did not discuss the "why" associated with the results of his and Moreno and Mayer's (2000) studies. More recent texts, such as Gibbons' *Multimodality, Cognition, and Experimental Literature* (2012), integrate discussions of neuroscience attributes of multimodality, particularly from linguistic or psychological disciplinary perspectives, however, they do not integrate the language of neurobiology into those discussions. Studies in neurobiology related to cognitive neuroscience shed light on why certain modal combinations, particularly those that include visuals, tend to work well in learning environments and can contribute to theorizing the rhetoric of multimodality.

It is necessary to integrate neurobiological terminology to explain why certain multimodal combinations and rhetoric works. I review some of this literature in the remainder of this chapter. I avoid thick description of these studies, and I provide definitions of several terms in order to provide a primer for interdisciplinary theorization. For example, while I describe relevant concepts such as "mirror neurons" (to which I alluded much earlier in this chapter), I do not refer to specialized names given to neuron types such as "F5 neurons." One does not need to be a biologist to understand findings related to bioneurological and neurobiological studies and their connection with multimodal rhetoric, however, one needs to be able to understand and use some terms in order to appreciate the discourse of the discipline from which the analysis comes.

In this overview, I also identify specific theories of neurobiology that I use in my own theorization connecting these disciplines. Four specific concepts connected to neurobiology are involved with this effort: (a) the already-mentioned technologies related to collecting data, which integrates some discussion of affordances and constraints of technology—a theory recognized by many rhetoric and composition scholars; (b) multimodality of neurons; (c) mirror neurons; and (d) neural plasticity. Further, this theorization integrates neurobiological perspectives of multimodal rhetoric, including the Colavita visual dominance effect. Rhetoric scholars recognize the role visual rhetoric plays, and it has been theorized considerably, however, none of it has integrated the notion of the Colavita visual dominance effect. In addition to their interest in dynamics related to visual information, neurophysiologists also recognize the important role of previous experience and knowledge in cognition as well as how information is presented relative to modes involved and timing of presentation, similar to the findings associated with Moreno and Mayer (2000). These are integrated into the model that I present. However, I provide some details of the three neurobiological theories that serve as the foundation of the biological attributes of the discussion next.

NEURAL RESEARCH METHODS

Data collection facilitates careful analysis toward theorization. The means by which data collection is facilitated plays a very important role in the nature of the data used. Neurophysiologists and neurobiologists tend to use three different methods that involve different technologies. Sometimes these studies will integrate more than one approach in an effort to triangulate data. I describe these three approaches here to provide an understanding of how data is collected within the field of neurobiology.

Electrophysiology

A popular approach to study neural activity is through studying electrical waves sent across the brain as one performs a particular task. Electrodes are placed on various parts of the head, and they monitor signals suggesting certain neural activity. Such technology is affected by the tool's capacity to measure multiple neurons and its range covering the area of the brain. Older technology was limited to studying very few neurons, and electrodes were most sensitive to activity near where they were placed. So they could not monitor much activity.

More recently, technology has enabled researchers to study more neurons and a larger area of the brain. Consequently, it is now possible to have a better understanding of what is happening with various neurons as one performs a task. However, it is still uncertain as to what activity is specifically occurring at the cell level.

The 2-Photon Microscope

As indicated above, powerful microscopes are able to provide an image of the brain as one performs a given task. The 2-photon microscope allows researchers to view activity, however, this is generally limited to study of smaller animals given the size of the microscope and radiation. However, it is possible to theorize human neurobiology from certain small animals such as rats. A constraint of the 2-photon microscope, though, is that it is limited to studying tissue closer to the brain surface than most neural activity occurs.

Hemo-Neural Hypothesis

It is generally held that neurons in the brain facilitate many cognitive processes. Neurons help to transport information from one part of the brain, as it is acquired, to other parts of the brain, where it may be processed. It is generally recognized that this is an electrical process, making electrophysiology a valid approach to studying the brain processes; this is what forms the foundation of computational theories of the mind. The brain is like a computer. Several studies note the relationship between neural disorders and cognition.

Also, the brain includes a vast system of vessels carrying blood to various parts of the brain. Indeed, as Gross (1998) pointed out, Aristotle believed the manifestation of the activity of this system of veins to be emotion, intelligence, and action (p. 247). Moore and Cao (2008) noted that blood flow in the brain "is typically well correlated with neural activity" (para. 45). As such, they surmised that blood flow can be an indicator of neural activity, citing several attributes of computational theory and intelligence derived from the Turing Test (Turing, 1950). Moore and Cao stated that,

> In many cognitive paradigms, blood flow modulation occurs in anticipation of or independent of the receipt of sensory input. One example of a context in which hemo-neural modulation of cortical dynamics may impact information processing is through enhancement of evoked responses during selective attention. A wide variety of studies has shown that attention to a region of input space (e.g., a retinotopic position or body area) is correlated with enhanced evoked action potential firing of cortical neurons with receptive fields overlapping the attended region (Bichot & Desimone, 2006). These effects typically emerge 100–500 ms after the onset of attentional focus (Khayat et al., 2006; Khoe et al., 2006; Worden et al., 2000). (para. 45)

In short, Moore and Cao theorized that the connection between neural activity and blood flow is so closely correlated that the two suggest a relationship by which neural activity can be measured by blood flow to certain parts of the brain relative to performance of particular tasks. For example, in highly visual task-oriented studies, a larger amount of blood is observed to flow to the visual (or occipital) cortex than otherwise observed. As such, it is possible to use blood flow—hemodynamics—to study neural activity and cognitive processes. Blood flow can facilitate analysis of neural inhibition (neurons being subdued or prevented from activity) and neural excitation (neural activity increasing). They indicate that fMRI technology allows researchers to observe and measure blood flow and thus to infer neurological processes *in vivo*, important to ascertaining such correlation and analysis (Cao, 2011).

Such findings of correlation between neural activity and blood flow suggest that multiple biological systems are at work within cognition. Much as multimodal rhetoric research has discovered that presenting information using multiple modes may affect cognition better than providing it within a single mode, multiple systems within the brain may be involved in processing information or facilitating such information processing. Rather than one system doing all the work, multiple systems contribute to cognitive processes.

Several studies use this hemo-neural hypothesis to theorize neural processes related to cognition and others use it to triangulate data from EEG methods (see collections edited by Calvert et al., 2004 and Murray & Wallace, 2012). Further, these studies assert conclusions consistent with studies of multimodal rhetoric, explaining some of the biology behind those conclusions.

The topic of technology has been part of rhetoric and composition scholarship for several years now. Integrating neurobiological studies into the study of the rhetoric of technology opens the door to further analyses of how technologies influence research and contribute to new knowledge about cognitive processes and ways to present information productively to facilitate cognition. Neurobiologists, for example, have studied the ways that simulators affect neural processes, examining blood flow across the brain as subjects interacted with a simulator. Multimodal rhetoric scholars have studied the use of simulators in learning by having subjects perform certain tasks after interacting with a simulator, however, they do not integrate a discussion of neural processes involved in that interaction. Combining research methods and analyses from both fields into a single study may enhance such studies.

Multimodality of Neurons

There are two kinds of neurons identified in the neurobiological literature relative to modal attributes: unimodal and multimodal. Unimodal neurons carry information relative to a single modality. For example, a unimodal auditory neuron can process only auditory information. Multimodal neurons are neurons that can carry information relative to more than one mode. A multimodal neuron may be able to process information from both visual and auditory senses (multimodal). A neuron that can carry information relative to three different modes is particularly labeled as trimodal, however, it can be discussed within the multimodal neuron category.

Again, older technologies tended to focus on unimodal neuron activity because of their limited capacity; but more recent technology is able to show multimodality of neurons. There is no debate about the existence of these modal characteristics, however, there is some debate regarding whether particular sets of neurons act independently of one another and can manage facilitating individual sensory information and then bring it together computationally, or if they can facilitate multiple sensory information (Allman, Keniston, & Meredith, 2009; Allman & Meredith, 2007; Bernstein, Auer, & Moore, 2004).

Studies related to multimodal neurons also suggest that certain kinds of information can be processed at different rates by such neurons, suggesting an optimal modal composition of a given message to facilitate faster processing (Bethge, Rotermund, & Pawelzik, 2003; Bremner & Spence, 2008). Numerous studies in multimodal rhetoric examine combinations that have the best effect on an audience (Kress, 2009, Moreno & Mayer, 2000). Studies in assessment of multimodal products contribute to this analysis as well (Ball, 2006; Odell & Katz, 2012, Neal, 2011; Remley, 2012). As such, multimodal rhetoric scholars can contribute to studies related to ascertaining optimal combinations by developing potential multimodal products that can be used in empirical electrophysiological or fMRI studies.

Mirror Neurons

Gallese, Eagle, and Migone (2007) and Rizzolatti, Fadiga, Fogassi, and Gallese (1996) first reported on the existence of neurons that appear to facilitate cognition of movements and behaviors that one observes another perform while doing a given task. Even before the observer tries to perform the same task he or she observed, he or she has acquired a sense of how to perform the task through a mental visual mirror. Further, they observe a connection between these neurobiological phenomena and social science. They state,

> Suppose one sees someone else grasping a cup. Mirror neurons for grasping will most likely be activated in the observer's brain. The direct matching between the observed action and its motor representation in the observer's brain, however, can tell us only what the action is (it's a grasp) and not why the action occurred. This has led some authors to argue against the relevance of mirror neurons for social cognition and, in particular, for determining the social and communicative intentions of others (see, e.g., Jacob & Jeannerod, 2004; Csibra, 2004). (p. 135)

Mirror neurons facilitate much of the cognition associated with experiential learning and hands-on training. One can learn a certain amount by reading about it, but the adage "experience is the best teacher" still is generally accepted as true. Observation and action are that experience, and mirror neurons help actual learning associated with those activities. This echoes scholarship in multimodal rhetoric applied to experiential learning.

Neural Plasticity

In cognitive neuroscience, plasticity pertains to the ability of neurons to change their composition and behaviors relative to the information they process and experiences. Neurobiologists recognize, much as humanities scholars such as Gee, Pinker, and Mayer, that experience plays a role in learning. What one understands of a given bit of information and how they tend to best learn information affects how they learn new information. Berlucchi and Buchtel (2009) define neural plasticity as

> changes in neural organization which may account for various forms of behavioral modifiability, either short-lasting or enduring, including maturation, adaptation to a mutable environment, specific and unspecific kinds of learning, and compensatory adjustments in response to functional losses from aging or brain damage. (p. 307)

Studies related to plasticity tend to examine how one responds to a series of subsequent experiences of certain modal combinations after first exposure, especially related to cognitive development. Generally, the brain is able to process information more quickly as it learns more about that information. Depending on the amount of exposure to information and the way the information is provided, cognition about the information can occur more quickly.

Multimodal rhetoric scholarship in learning examines similar attributes, that is, how one learns new information is affected by their background and learning methods, including modal combinations (e.g., Moreno & Mayer, 2000). Much scholarship in rhetoric has looked into virtual worlds and their benefits in cognition and learning (e.g., Gee, 2003; Remley, 2010, 2012; Ritke-Jones, 2010; Vie, 2008). A growing body of scholarship in biological neuroscience also considers the role virtual environments play in learning and literacy development (Donahue, Woldorff, & Mitroff, 2010; Li, D'Souza, & Yunfei, 2011).

Another area of plasticity study is the effect of brain injuries or disorders on cognition. Some scholarship in rhetoric theorizes relative to such disorders and injuries. For example, there is some study in the fields of rhetoric and cognitive psychology about how to facilitate learning among those with autism. Like studies I have described already, these tend to involve providing a subject with some instructional material and then assessing performance on a given task related to that material to measure learning. Neurobiologists try to examine neural processes as one reviews instructional material to understand what modal combinations may facilitate certain kinds of neural activity involved in learning. So it is possible to merge the study of rhetoric and the effect of brain disorders/injuries on cognition.

Because of this body of literature that overlaps much of what has been done in multimodal rhetoric scholarship and the growing interest in interdisciplinary study, it is important for disciplines to find ways to synthesize their work. The disciplinary divide related to discourse differences discourages such interdisciplinarity, and scholars in both areas recognize how this negatively impacts the study of cognition. However, this text attempts to build a bridge by which rhetoric and social science scholars can contribute to biomedical cognitive neuroscientists and consequently recognize the value of rhetoric scholarship in STEM programs and vice versa. I outline the remainder of the book in the following section.

OVERVIEW OF BOOK

In the second chapter, I review extant literature that I have mentioned in this chapter to detail connections between the science-oriented fields of cognitive neuroscience and multimodal rhetoric. I connect multimodal rhetoric theory with perspectives about multisensory processing from neurobiology. This review also facilitates identification of some attributes of multimodality that contribute to an integrated theory of multimodality; this theory integrates elements of biophysiology and rhetoric.

In Chapters 3 and 4, I detail the new theoretical model, and I provide analyses of cases related to it in Chapters 5–8. Such analyses provide illustrations of possible ways to apply the new model to analyze multimodal rhetoric and neuroscience via an interdisciplinary approach. Chapter 5 in particular features a detailed case study related the Training Within Industry (TWI) program as it is practiced in business and industry. The program favors hands-on training using a particular approach. It was initiated during World War II to address low literacy levels in workers who were making a transition from farmwork to war-industry work. In this discussion, I provide some background related to the TWI program and the development of its rhetoric, including the ineffective applications that contributed to a catastrophic accident prompting changes toward more effective applications. I build on existing multimodal learning theory (especially that of Moreno & Mayer, 2000, and Tufte, 2006), and I introduce new concepts of visual rhetoric relative to scholarship in cognitive neuroscience and how that scholarship theorizes interconnections of multisensory stimuli. In Chapters 5 through 8, I apply the model to specific case studies integrating different forms of new media used in training and education.

In Chapter 9, I discuss some implications related to the new theory as it applies to education and workplace training and interdisciplinary research. Finally, in Chapter 10, I argue for further study into the neuroscience of learning and multimodal designs that facilitate learning and for a closer connection between neuroscience and multimodal rhetoric scholarship.

CHAPTER 2

Multimodality and Neurobiology

One of the goals of this book is to raise awareness that neural activity affects how different modal combinations may work best for certain purposes and certain audiences or individuals. Studies like Moreno and Mayer's (2000) brought attention to the issue of optimal modal combinations for learning, but we do not know why those combinations worked well in that particular set of experiments. Neuroscience is the study of how neurons in various parts of the brain act under various stimuli. Cognitive neuroscience focuses study on these actions relative to learning. That is, it considers how the brain responds to various stimuli when the brain is tasked with a learning activity. The field of cognitive neuroscience integrates several fields, including cognitive psychology, philosophy, chemistry, and biology.

In this chapter, I discuss the link between multimodal rhetorical theory and neurobiological scholarship in cognitive neuroscience toward identifying why multimodal materials facilitate learning. I also hope to encourage scholarship that links these fields further; I identify potential avenues for studying this link at the end of this chapter as well as in the book's conclusion. In this chapter, in particular, I review scholarship related to cognitive neuroscience and its multisensory attributes. I eliminate biological jargon to focus attention on the general link between the fields of neuroscience and multimodal rhetoric. This information sheds light on why people tend to learn better when multiple modes are used to present information. Limiting the terminology to a small set of new vocabulary will also help transition rhetoricians to the science-oriented language.

Little scholarship in multimodal rhetoric uses the scholarship from the field of cognitive neuroscience. One reason for this is that the biology associated with neuroscience quickly becomes too complex for most people to understand. It includes thick discussions of neurons and how information travels around the brain, integrating very specialized language associated with the biology of the brain. I try to describe some related scientific concepts without using that jargon.

The scientific fields—physics, chemistry, biology, and related subfields—tend to focus study on neuron behaviors. Neurons are often considered as a network of information-processing paths. More specifically, neurons have three principle parts: axons, dendrites, and cell body (Figure 2.1). The junction between two neurons is called the synapse. Dendrites take sensory information in, and axons send information out. The neural process involves a signal being sent from the axon to the dendrite via synapses. However, the brain develops new neurons in developmental stages of life, and it kills off neurons as we mature and grow old.

Functional magnetic resonance imaging (fMRI) is one of the newest technologies used to observe neural activity in the brain as one performs a task. In particular, as one performs a given task or acquires information, certain parts of the brain are shown in fMRI scans to be more active than other areas of the brain relative to blood flow. One of the more interesting, novel studies of blood flow and task process was the study of blood flow as one views pornography. Huynh et al. (2012) found that blood flow to the visual cortex while viewing pornography (high-intensity, visual erotic stimuli) is less than normal relative to neutral visual stimuli, however, blood flow to other parts of the brain associated with sexual performance/arousal were activated.

Analysis of hemodynamics and electrophysiology pertaining to neurons helps researchers understand which parts of the brain are working most during certain kinds of activities. Different cortices are associated with different functions. Figure 2.2 shows the four cortices (or lobes) of the brain.

The frontal lobe is generally associated with language skills, reasoning, high-level cognition, and motor skills. The parietal lobe is associated with touch (hapo-stimuli) and memory management of other senses. If damage to this lobe

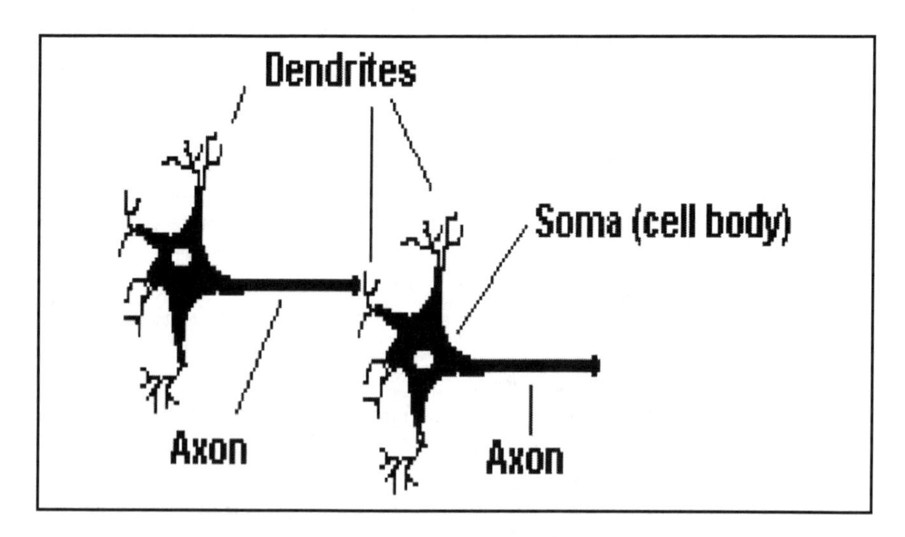

Figure 2.1. Parts of a neuron.
From: http://faculty.washington.edu/chudler/synapse.html

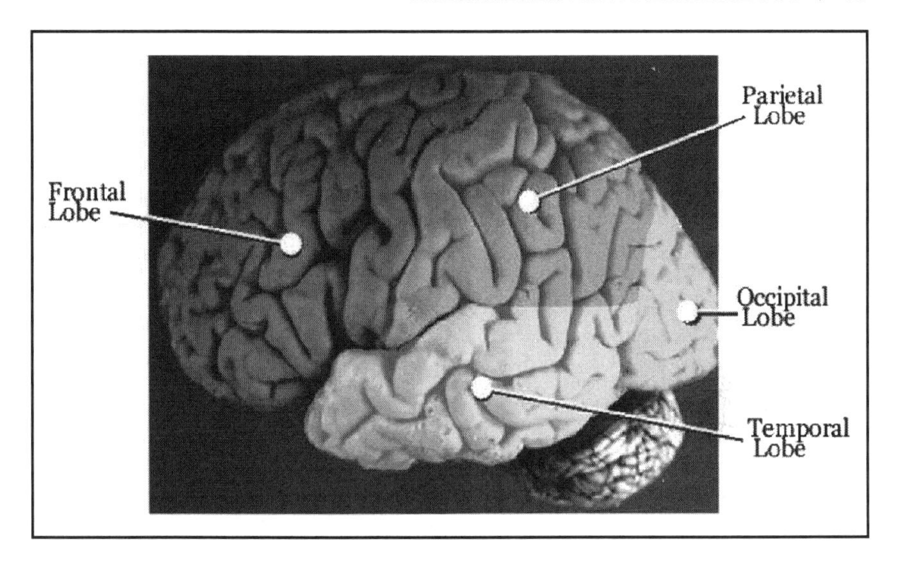

Figure 2.2. Cortices of the brain.
Image by John A. Beal, Louisiana State University Health Sciences Center Shreveport.

occurs, one can have difficulty, for example, with recalling how to speak correctly. The occipital lobe is associated with visual stimuli (visual cortex), and the temporal lobe is associated with sound (auditory cortex) and memory associated with certain kinds of sensory information.

As more blood moves to a given cortex of the brain, the implication is that the particular neural functions of that cortex are more actively engaged than are other parts of the brain; this is referred to as the "hemo-neural hypothesis" (Moore & Cao, 2008). Tools such as fMRI help identify patterns of neural activity and blood flow. Most of the research prior to 2000 focused on particular individual modal dynamics such as how the brain operates when stimulated by a single particular mode (e.g., hearing, vision, or touch). However, advances in MRI technology allow closer examinations, and recent studies have found that more than one sense is generally engaged in various activities, but to varying degrees based on the activity.

There is considerable literature in the neurobiological field of cognitive neuroscience that characterizes learning as a multisensory experience (see collections edited by Calvert et al., 2004 and Murray & Wallace, 2012). Each sense integrates a particular mode that has already been identified in previous chapters: touch, spatial presence, visual perception, and auditory. The various modes of representation are among the attributes of cognition related to learning, and the field of neuroscience asserts that cognition is a multisensory experience, which encourages use of multimodal instructional materials. Work by Moreno and Mayer (2000) and Mayer (2005), which I have already identified, described links between learning and multimodal instructional materials.

One sense not identified with linguistics and literacy that is also included in the scope of cognitive neuroscience is smell. I integrate this sense into this discussion on a limited basis, because it contributes to experience and learning. When one watches a television show about lions or penguins or other wild animals, he or she can observe animal behaviors and, depending on the quality of the microphones used, hear the sounds the animals make as they communicate with each other. A similar experience occurs when one goes to the zoo. However, the biggest difference between watching a television show about animals and going to the zoo is that one cannot smell the animals through the television experience. One is able to smell the animals when they go to the zoo. While that part of the experience may be among the less pleasant sensory experiences, it contributes to that particular kind of experience and may be part of learning about animals. For animals, distinguishing smells is among the necessary skills for survival. Animals learn to smell other animals—predators—and use that information for survival, being more careful around or avoiding smells associated with predators. That may be less important for humans, but the workplace surroundings and smell may contribute to the learning that one experiences there. Indeed, Krishna, Elder, and Caldara (2010) linked touch and smell to consumer behaviors, indicating that smell plays a role in purchasing behavior. While smell is included in this discussion, a majority of it concerns the other modes and related senses already identified.

Multimodal Integration and Cognition

As I mentioned in Chapter 1, recognizing that multiple senses are involved in cognition, neurobiologists refer to multisensory processes as "multisensory integration," or "multimodal integration." This terminology helps facilitate a connection with multimodal rhetoric. However, I want first to review very generally how multimodality is treated in the field of rhetoric. This sets up a more detailed discussion of that literature and connections with neurobiology.

Many studies of literate practices have discussed multimodality more explicitly (Gee, 2003; Lemke, 1998, 1999; Mayer, 2005; Moreno & Mayer, 2000; New London Group, 1996; Richards, 2003; Whithaus, 2012). In a seminal theoretical piece, the New London Group (1996), which is composed of several literacy scholars, observed that new media forms affected ways by which people could make meaning and various practices that come into play in various contexts. They identified five different, unique modes of representation: print-linguistic, visual, audio, gestural, and spatial; and they acknowledged that any two or more of these can be combined to form a multimodal representation. They acknowledged that,

> We argue that literacy pedagogy now must account for the burgeoning variety of text forms associated with information and multimedia technologies. This includes understanding and competent control of representational forms that are becoming increasingly significant in the overall communications environment, such as visual images and their relationship to the written word—for instance, visual design in desktop publishing or the interface of visual and linguistic meaning in multimedia. Indeed, . . . ; the proliferation of communications channels and media supports and extends cultural and subcultural diversity. (p. 60)

With this observation, researchers began considering various combinations of modes of representation that can contribute to communication. Further, scholars realize that pedagogy needs to integrate instruction in composing in these different modes of representation. The importance of graphic images in these literate practices is noteworthy because of the different kind of literacy at work relative to each—print-linguistic text and image, though both represent communication systems (J. Murray, 2009).

In the past 15 years, another focus of study within literacy studies has emerged that focuses on the use of multiple modes to communicate and related practices. Studies pertaining to this analysis seek to understand rhetorical attributes of mixed modes and when and under what conditions certain combinations are most productive (e.g., Lemke, 1998, 1999; Richards, 2003). J. Murray (2009) indicated that because of connections between language and consciousness, a given combination may be meaningful for some people while the same combination will not be as productive for others because of differing backgrounds, which include not just literacy training but literacy experiences and understanding of the universe (J. Murray, 2009, p. 16). Indeed, several theories of multimodality and this relationship between print-linguistic text and image have been presented, but each seems to have its own difficulties meeting the challenges of theory development thereof.

Kress and Van Leeuwen (2001) attempted to develop a theory of semiotics that integrates terminology describing various rhetorical dynamics at work in multimodal forms of communication. They recognized the relationship between composer and "reader" as an interactive one: The reader acts upon a message as much as a composer initiates it. Furthermore, Kress and Van Leeuwen articulated that the message is not just the content but its form relative to how it is presented and the communicators' relationship to each other and their experiences. The terminology they used includes discourse, design, production, and distribution (pp. 4–5). Discourse pertains to socially constructed knowledge. Design pertains to the resources that one uses to create meaning. A diagram of an object represents a different kind of visual representation than does a photograph of the same object: A diagram will integrate labels and dimension information, providing details that a photograph may not be able to provide. Design frames how meaning is made. Production describes the materiality of the expression, the media used to make the message visible/material. A print-linguistic document is able to provide information a certain way, however, a video demonstration presents the same information in another way, while a real-time, in-person demonstration facilitates information yet a third way. Distribution, similar to the fifth rhetorical canon of delivery, refers to the means by which readers access the material message (pp. 20–21). Multimodal presentation is facilitated through a system that affects meaning-making: how people of a given community understand language (discourse), how people understand certain conventions of communication (design), how people use certain tools to develop a given message or artifact of communication (production), and how people disseminate those artifacts, all of which are affected by available technologies.

Kress and Van Leeuwen (2001) also applied certain terms to explain the social interaction involved in making meaning. "Mode" pertains to the different "genres" of composition identified by the New London Group (1969), of which Kress is a member. These include aural, visual, print-linguistic, experiential, and spatial as well as combinations. The term "medium" pertains to the material resources used to produce a message. "Experiential meaning potential" is a term that pertains to the roles that people's past experiences play in contributing to making meaning out of a given message. One who has only learned how to do new tasks through demonstration and practice and has never used a print-linguistic document will not understand how to use a manual to learn a new process. Finally, "provenance" pertains to the importing of certain signs into different contexts to help people understand each other's ideas and values (pp. 21–23).

Much of the literature in cognitive neuroscience pertains to the integration of different sensory experiences toward facilitating learning and cognition—how neurons react to stimuli and how different modes of representation and different sensory experiences related to those modes interact with each other neurologically to affect learning and cognition. While some studies focus on questions pertaining to development of particular neuroprocessing combinations and how different modalities are processed, some studies consider the development of this integration itself. Much of this book revolves around the studies pertaining to how neurons interact around the brain during certain tasks, not on the development of this integration. However, I want to point out that the Gestalt effect is identified in much of the scholarship on visual rhetoric (especially, Arnheim, 1969), yet it is also included as a term/concept in the neuroscience scholarship (Wallace, 2004). Wallace (2004) stated broadly that one of the roles of the brain "is to synthesize this mélange of sensory information into an adaptive and coherent perceptual Gestalt . . . this sensory synthesis is a constantly occurring phenomenon that is continually shaping our view of the world" (p. 625). I review some of the literature from neurobiology regarding various modal combinations and how they are described in that scholarship. This discussion omits studies pertaining to demonstrating cognition, which involves another dynamic of neural process. For example, studies have found that the neural activity related to deciding to do a particular action precedes the display of that action (Massumi, 2002).

Bremner and Spence (2008) found that multisensory experiences have a more positive effect on learning than do single-sensory experiences, suggesting that learning is better facilitated through multimodal instructional materials. Lewkowicz and Kraebel (2004) asserted that redundancy of different sensory experiences helps to reinforce information and understanding of that information. Keetels and Vroomen (2012) observe that access to different features across different senses "increases perceptual reliability and salience," improving opportunities for learning and speeding up the reaction to stimuli (p. 147). As an optimal combination of senses is engaged, the brain is able to process the information faster. They study how the neurons process information relative to the temporal synchronicity of the sensory experiences. That is, they consider the amount of time that passes between receiving visual information and auditory information or other modes engaging different senses.

While the general understanding is that the less time that passes between the two senses are engaged simultaneously the better able the brain is to process both bits of information, Keetels and Vroomen (2012) question the time needed to process different combinations and the rate at which each mode or sense can be engaged optimally. Just as sound travels slower than light, they acknowledge that, "the neural processing time also differs between the senses, and it is typically slower for visual than for auditory stimuli" (p. 148). So, while Moreno and Mayer (2000) found that the narration and visual information need to be positioned closely, such positioning of other modal combinations may not affect learning similarly.

Demonstrating a task engages visual and auditory senses, however, practicing the task immediately after the demonstration engages visual, auditory, as well as touch and spatial senses. Such experiences may facilitate certain kinds of learning better than other combinations. The rest of this chapter summarizes attributes of particular modal combinations relative to cognitive neuroscience.

NEURORHETORIC OF PARTICULAR SENSES

Visual-Dominance Effect

There is much literature about visual rhetoric. Visual representations occur in many forms and are prevalent. Indeed, Arnheim (1969), Mitchell (1995), and Pinker (1997) argued for examination of visual representations and how the mind processes them. Images are important to one's cognitive understanding of abstract ideas and concepts and the relationship of image to text (Arnheim, 1969; Mitchell, 1995). While Arnheim valued image over words, Mitchell advocated equality between the two, suggesting that value of one over the other depends on context. As many scholars have noted, a reader who is given a visual representation of a given process is able to understand better what the process involves than one who has read print-linguistic text exclusively (Arnheim, 1969; Mitchell, 1995).

Arnheim (1969) was among the earliest scholars to argue for the value of images to facilitate cognition. While print-linguistic text had been emphasized generally (as also evidenced in the historical study), visuals carry with them what Arnheim called a "Gestalt effect." Visuals can be seen in their entirety, while words must be processed individually, breaking up flow of information. He explained that the mind creates and uses any visually represented information to understand the world, yet the visual is traditionally treated as subordinate to words and numbers (p. 308).

Arnheim's (1969) main point was that in order to best accommodate cognition through visual perception, pictures need to articulate meaning explicitly or link closely to the reader's prior experiences to facilitate meaning-making: "Pictures and films will be aids only if they meet the requirements of visual thinking" (p. 308).

Hicks (1973) also called attention to a relationship that exists between words and images. Summarizing some main principles of photojournalism, he acknowledged that photojournalism is all about combining words and pictures or the "coming together of verbal and visual mediums" (p. 3). He explained that the verbal medium is discursive (has a set way of being read), while visual forms can be read any number of ways. He further stated that a perfect relationship between visual and verbal cannot be attained. With the right words, the visual can perform its function better.

Mitchell (1996) also observed that images can be read various ways, and they interact with words various ways as well. He alluded to the concept of *ekphrasis*—that one can state a given message using various combinations of image and text together. The concept of multimodal rhetoric suggests that authors of multimodal works need to try to find ways to combine the various modes available—not just words and images, but possibly audio and space among others—to accomplish a given purpose most effectively and efficiently.

Mayer (2001) identified three potential learning outcomes: no learning, rote learning, and meaningful learning. Ultimately, Mayer characterized multimedia instructional messages as communicating "using pictures and words intended to promote learning" (p. 21). One of Mayer's multimodal principles was that people learn better when pictures and words are integrated into an instructional message than when only words are used (p. 63). When only words are used, people may attempt to "build a visual model." However, they may not attempt to do so, or the visual model they construct is erroneous. If a picture is provided, people can make the visual connection more readily. Empirical evidence that supports this assertion exists (pp. 72–78).

Indeed, it is generally recognized in neuroscience scholarship that vision is the dominant modality in humans (Colavita, 1974; Howard & Templeton, 1966;

Welch & Warren, 1986). Spence, Parise, and Chen (2012) describe the neurological basis for this apparent favoring of the visual over other modes of representation. They review literature on the Colavita (1974) visual-dominance effect, which recognizes that visual stimuli tend to get a stronger response from neurons than other stimuli. Colavita found, in a series of four experiments, that subjects responded to visual stimuli before they responded to any other sensory stimulus (p. 411). Further, he noted that scholarship generally recognizes that people "do not respond as effectively to two simultaneously presented stimuli as to the same two stimuli presented in succession." (p. 412). However, Spence et al. (2012) suggest that this effect is attributable to the fact that the visual information is received generally before any other information—auditory or touch, for example (p. 537). Because vision is the first sense that receives information, it is naturally engaged immediately, though, auditory information is actually processed faster than visual information.

Also, some studies found that the visual sensory experience is prominent in learning and in the development of multimodal integration (Wallace, Perrault, Hairston, & Stein, 2004). Other studies suggested that the other senses do not adequately compensate for the absence of visual senses, for example, in people with low vision or blindness, negatively affecting learning abilities (Elmer, 2004; Pasqualotto & Proulx, 2012). Donahue et al. (2010) reported that video game players tend to have refined visual attention and visual perception abilities, contributing to enhanced multisensory processing ability.

Visual rhetoric has received a lot of attention, however, other senses come into play in cognition as well. Most of it recognizes the esteem given to visual attributes. Neuroscience recognizes the value of visual modes in facilitating cognition, however, it also finds that other senses contribute to the learning processes as well. One of the most studied combinations is the visual-auditory combination.

Visual-Auditory Links

Moreno and Mayer (2000) reported findings associated with several experiments they conducted toward developing a cognitive theory of multimodal learning. These experimental designs involved participants viewing multimodal texts and responding to particular questions about information in those texts to ascertain how much the participants learned. In one study, they reported that 78 college students viewed animation associated with the process of lightning generation combined with either narration describing major steps in lightning formation ("AN group") or with text on screen that involved the same words and timing ("AT group"). The groups were then tested on transferred knowledge, and the AN group developed more correct solutions than the AT group on the transfer of knowledge test. In another experiment, they report that 137 college students observed animation

> in one of the following six conditions: one group of students viewed concurrently on-screen text while viewing the animation (TT), a second group of students listened concurrently to a narration while viewing the animation (NN), a third group of students listened to a narration preceding the corresponding portion of the animation (NA), a fourth group listened to the narration following the animation (AN), a fifth group read the on-screen text preceding the animation (TA), and the sixth group read the on-screen text following the animation (AT). (para. 9)

The narration-related groups performed better than the text-related groups in problem-solving knowledge transfer (Moreno & Mayer, 2000, para. 10). In a third experiment, Moreno and Mayer (2000) reported that 69 college students participated in a two-way analysis of variance "with the between subjects factors being redundant or non-redundant verbal information (A-NT and ANT versus A-N and AN, respectively) and simultaneous or 'sequential presentation' order (AN and ANT versus A-N and A-NT, respectively)" (para. 13). Moreno and Mayer concluded that students who received "sequential presentations" performed better (more creative solutions) on the knowledge-transfer test (para. 13). Moreno and Mayer's work with these experiments is considered seminal in multimodal/multimedia instructional theory, however, they never mention Colavita (1974) in their reports of the experiments or theorization.

Generally, Moreno and Mayer (2000) found that certain combinations of visuals and text information affect learning, suggesting a relationship between modes used to communicate and their rhetorical impact. In an instructional context, combining visual and verbal/aural modes of representation is more powerful for accomplishing the instructional purpose than using only narration or visuals alone. From these experiments, they formulated several principles associated with multimedia instruction. Mayer (2001) summarized their multimodal principle with the statement that people learn better when pictures and words are integrated into an instructional message than when only words are used (p. 63). When only words are used, people may attempt to "build a visual model," but they may not attempt to do so. If a picture is provided, people can make the visual connection more readily. Mayer also asserted that it is vital to eliminate extraneous material—words, images, and sounds—from any multimedia message. Such irrelevant information "competes for cognitive resources in working memory," disrupting the learner's ability to organize and retain relevant information (p. 113).

The auditory cortex is the portion of the brain where most of our auditory (hearing) processing occurs. Neuroscience scholarship explains this benefit in the linkage of visual stimuli being processed within the auditory cortex. According to Bizley and King (2012), "more than one quarter of the neurons associated with the auditory cortex are influenced by visual stimuli" (p. 37). They go on to state that visual stimuli enhance the processing that occurs in the auditory cortex. In particular, they state that visual inputs can increase the sensitivity and selectivity of responses to auditory information (p. 40). This is supported in other studies as well (Kajikawa, Falchier, Musacchia, Lakatos, & Schroeder, 2012; Kayser, Petkov, Remedios, & Logothetis, 2012; Munhall & Vatikiotis-Bateson, 2004; Newell, 2004). Newell (2004) suggested that auditory information complements visual information to help refine one's understanding of the information. Likewise, visual information may facilitate refinement of auditory information, as when one hears a siren from afar and then sees a fire truck (as opposed to a police car). The visual of the fire truck helps the person understand that a fire-related emergency is occurring. Linguistic scholarship generally observes that one benefits from facial expressions (a visual stimulus) as nonverbal cues when one is speaking (auditory) to them. Bernstein et al. (2004), though, identified some debate regarding whether this relationship is one of convergence (facilitated by multimodal neurons—information-processing occurring at same time with same neurons) or association (neurons of two different modalities being used to process different kinds of information). They concluded that it is one of association (p. 218).

However, scholarship in neuroscience also indicates that the auditory sensory experience can affect visual perception related to other modalities, including space, time, and motion (Shams, Kamitani, & Shimojo, 2004). So, while one experiences visual information, other senses can affect how that information is perceived.

The Rhetoric of Hands-On Learning

A growing movement in education is called experiential learning, which encourages students to learn skills through performance of them—hands-on training, in workplace terminology. It encourages less reading and lecture-related instruction in the classroom and more practice with the actual tasks in a real or realistic environment. Such a philosophy recognizes that cognition occurs by doing the task rather than by merely reading about it. The various modes of representation involved in such a learning setting and the ways to present them are of interest to multimodal rhetoric scholars as well as neurobiologists.

One of the earliest studies related to multimodal rhetoric was actually focused on manuals and performance. In the 1980s, a debate in the field of workplace literacy practices emerged concerning the use of comprehensive manuals versus what was called a "minimalist manual"—a manual that limited information to the particular steps involved and omitting background and theoretical information. However, the study also involved a group of people who learned to perform a task by doing it rather than by reading a book about that task.

Carroll, Smith-Kerker, Ford, and Mazur (1988) used two experimental studies to measure efficient learning with different manuals. The first study used a between-subjects design in which 19 participants used either a minimalist manual or a commercially prepared self-instructional manual to perform eight different word-processing tasks. Minimalist manuals remove much textual information that explains why certain steps are included in favor of focusing on performance steps. Performance was measured relative to length of time to learn a task and the success participants had doing the tasks. Participants using the minimalist manual were better able to complete the tasks more quickly and with more success than those using the self-instruction manual. In a second experiment, they used a similar design to measure performance relative to the 32 participants' use of either a self-instruction manual or minimalist manual and learning while doing or learning by reading the book. Participants were separated into four groups: two using the minimalist manual, one of which also learned the task by reading the book and the other that learned by doing the tasks; and two that used the self-instructional manual, one of which learned by reading the book and the other learning by doing. The "learn-by-doing" group was given 5 hours to complete a series of six tasks, while the "learn-by-reading the book" group was given 3 hours to use the manual to learn about the tasks and 2 hours to perform the six tasks. Generally, the results showed that the group using the minimalist manual was able to perform better than the groups using the self-instructional manual in both cases.

Also of interest is that in all cases, the group that learned by doing the tasks performed better than the group that learned the tasks by reading the more extensive book about the activity. For example, the group that read the minimalist manual and learned by doing the tasks had better comprehension scores than the other groups (23.8 subtasks completed successfully versus 20.4 successfully completed subtasks for the group that used the minimalist manual and read the book), though they took a bit longer to perform those tasks (121.7 min. compared to 71.8 min. for the group using the minimalist manual and the book). However, this group also experienced a few more errors in the tasks and took a bit longer to recover from those errors than the group using the minimalist manual and learning by reading the book (188.8 overall errors versus 187.3 overall errors; and 153.3 min. to recover versus 121.7 min., respectively). So, while people may learn more of a task by doing it, they tend also to take more time to perform it and have more errors than those who learn "by the book."

Charney, Reder, and Wells (1988) used experimental designs in three studies. The first study reported 40 inexperienced computer users grouped into one of two groups: One group was told which tasks they would be asked to perform using a PC-DOS system, and the other group was not informed of particular tasks they would be asked to perform. Half of each group was given one of two different manuals explaining basic commands for a PC-DOS system: One manual integrated more descriptive language, definitions, examples, overviews, and analogies than the other, which was about one third as long as the more elaborate version. After allowing the participants 45 minutes to read their manual, the participants were asked to perform various basic tasks explained in the manuals. The researchers measured how well participants performed the tasks relative to the number of tasks they could complete and how long it took to complete those tasks. They found that the group that was given advanced information about the tasks they would perform did better using the abbreviated manual than those who used the longer manual, and that those who had no prior information about the tasks did better using the longer manual than those who used the shorter manual.

Charney et al.'s (1988) second experiment reported on how well users responded to four different sets of manuals. They found that relative to degree and type of elaboration, there was no benefit from elaborations of general concepts, no benefit from elaborations offering advice of when to apply specific procedures, but significant benefit from situation contexts for applying procedures. The third experiment reported 30 participants studying VisiCalc commands by reading a particular manual at their own pace and then were asked to solve problems presented in the manual. Each participant was given three different problem sets: a tutorial acknowledging specific steps to use to perform a task, an exercise identifying a specific goal but no information on how to perform it, and a combination of tutorial and exercises. Performance was measured relative to percentage of task completed correctly and the amount of time needed to complete the task. Results showed that participants tended to perform better on tasks when there was an exercise related to the activity. With this study, they found that, relative to users and goals, if one has a specific task in mind, there is no need for elaboration, whereas readers with no specific task in mind benefited from some elaboration. Generally, these studies, like others I review, found that the "minimalist manual" facilitates more efficient learning progress than a self-instruction manual. However, they also suggest that the amount of prior information about the task people are given also influences how well they learn a related task.

In the next sections, I describe how neurobiology scholarship treats the various components identified with hands-on learning, including more specifically those related to visual, touch, spatial, and movement as well as previous experience. Much of it helps to explain the findings of the aforementioned studies, suggesting potential for a model that can bridge the fields to enhance scholarship and development of better instruction and instructional materials.

Visual-Spatial Links

Bizley and King (2012) observe that the visual processing that occurs in the auditory cortex seems to include visualization of spatial relationships especially related to sound location (p. 42). As one hears a given sound, neurons in the auditory cortex help one locate from where the sound is coming. Such information helps the listener focus on that location and carefully hear that particular sound. However, Burgess (2008) reported that different kinds of spatial information are processed by different sets of neurons, that is, information about different spatially related functions such as "consciously retrievable long-term knowledge" versus "'procedural' memory (e.g., habits, motor learning)" are processed by different sets of neurons (p. 92). Locating sounds is important in education and learning because it facilitates better attention.

Studies have also found a link between visual-spatial dynamics and motion perception.

Motion Perception

Moving about a particular place integrates several senses, including visual and vestibular (balance). These senses work to help one understand spatial movements, relative speed, direction of movement, and dynamic visual perception (changes in what the person sees as they move about). Neuroscience considers self-motion among its subjects of study (Campos & Bulthoff, 2012; DeAngelis & Angelaki, 2012; Soto-Franco & Valjamae, 2012; Wolbers, Hegarty, Büchel, & Loomis, 2008). One can learn spatial movements over a course of time; for example, one may learn to gauge where they are visually and spatially while driving to and from work each day. This information helps one monitor their progress when there is little light (driving at night or early morning).

Such information—understanding spatial relationships as we move about and interact in motion—is important because training involves learning patterns of behavior that include movement. As one practices a task, they move about a space, processing various sensory information prompted by the modes they encounter. As I have already mentioned here, different senses process information at different rates. Some training and interaction occurs in virtual environments, which integrate a virtual reality that differs in subtle ways from reality. Simulators and virtual worlds engage similar sensory experiences and certainly the same modes involved with self-motion perception. However, the digital representation modifies some elements of the interaction and sensory experience.

Campos and Bulthoff (2012), for example, point out that virtual reality technology modifies field of vision in a number of ways: One can see only what is available on a screen, while one's peripheral vision may be able to accommodate much more information. As I interact with Second Life, for example, that SL experience includes only the field occupied by a 17" screen, though my field of vision, including peripheral vision, extends to several objects to each side and above and below the screen. I am not immersed in the environment. I discuss some of the sensory processing problems related to certain kinds of simulations in a later chapter.

Touch-Vision

In demonstrations, one experiences visual and auditory information. When engaged with an activity, one experiences these senses along with touch (haptic).

Visuo-haptic processing helps one understand the spatial relations within objects as well as the form of objects. These enable one to distinguish parts of an object as well as distinguish between objects (Lacey & Sathian, 2012; Sathian, Prather, & Zhang, 2004). The way one understands an object is enhanced when one can touch that object (or tool) as opposed to just observing it being used (Newell, 2004). Lacey and Sathian (2012) call attention to this dynamic. However, they also state that one's background with a given object or task can affect how that person responds to similar objects or tools.

Previous Experience

Steven Pinker (1997) acknowledged that the mind works as a system that includes one's prior experiences and various forms of representation to understand information. Fields such as distributed cognition and social semiotics theorize the relationship between social interactions and cognitive processes. How one perceives the world is very much a function of the influence others may have on them. Teachers and parents exert considerable influence in shaping a child's perception of the world: Children learn how to think about the world by watching adults perform tasks and listening to someone tell them about the world. Any biases one has about the world can be transferred to a child whom he or she is teaching. Distinguishing artificial intelligence from human intelligence, Brooks and Stein (1994) noted that human "intelligence cannot be separated from subjective experience of a body" (p. 7). Even Schiappa (1993) argued the multiplicity of reality. What one perceives as "reality" is established through their interaction with the world. Such interaction involves experiences one has by himself as well as interactions with others that help to shape our understanding of our experiences. Two people who come from very different backgrounds may have different perceptions of the world, and this affects how they interpret information. Much scholarship in social semiotics and multimodality also observes this relationship, and scholarship in neurobiology and cognitive neuroscience also recognizes the role of previous experience in cognition. This study of the role of previous experience in cognition and learning is important because instruction that is similar to the ways a student or learner has learned other concepts will likely be more productive than an approach that is new to that learner. A model that recognizes and considers this role and one's prior learning experiences may facilitate development of more effective learning tools.

Relative to the concept of distributed cognition, Hutchins (2000) noted that "something special might be happening in systems of distributed processing, whether the processors are neurons, connectionist nodes, areas of a brain, whole persons, groups of persons, or groups of groups of persons" (p. 2). As systems interact, new knowledge is developed and cognition occurs. This knowledge development and cognition may be relative to the individual—neuron processing or to an entire culture—processing that occurs among groups of groups of persons, and any level in between. That is, distributed cognition can be at the level of the individual (distribution of systems within the brain) or at the level of culture (p. 5).

Hutchins (2000) also recalled Vygotsky's points that cognitive function occurs at two levels—interpsychological and intrapsychological. Drawing on Vygotsky and socially distributed cognition, Hutchins acknowledged that "the new functional system inside the child is brought into existence in the interaction of the child with others (typically adults) and with artifacts" (p. 5). Further, it is through these interactions that one can eventually learn to understand a situation "in the absence of the others" (p. 5). Neural plasticity facilitates this cognitive development within the individual's brain, but social interaction is a part of that neural development. Finally, Hutchins noted the contextual nature of cognition. He observed that "cognitive activity is sometimes situated in the material world in such a way that the environment is a computational medium" (p. 7). Learning occurs through experience. Part of learning is understanding how to respond to events similar to those that we have already experienced. That is, how did we respond initially? What result occurred? How can we change that result, if desired? As we work through an understanding of how to respond to such events or situations in the future, we can interact with others to find out how they reacted in such situations too. Hands-on training helps to provide a context for learning to do a given job or task. One performs actions with particular artifacts in the particular sequence they should be performed, and one interacts with others to better understand what they are doing and why.

Schnotz (2005) reviewed several studies pertaining to the influence working memory has on learning with multimedia, and developed a model of text/picture comprehension that considers working memory. Visual images that integrate text are easier to process because fewer processes of working memory are involved. According to Baddeley's (1986) model of working memory, there is a phonological (auditory) channel and a "visuo-spatial" (visual) channel associated with short-term memory. Based on his meta-analysis of empirical studies on students' learning ability with multimodal instructional texts, Schnotz found positive effects and negative effects of text/picture combinations. Among the positive effects were that coherence between text and pictures reinforce messages: When auditory and visual channels are used, the reader is able to process information better because attention is not split within a single channel; sequencing that places pictures before text are easier to process (pp. 60–61). Included among negative effects were redundancy; mapping may be complicated; information processing may be superficial, not deep; and some readers may use only certain representations and not others provided (pp. 62–64). Schnotz suggested that when a visual image is presented to a reader, the reader can create a visual model as he/she listens to a narrative about the picture. If only text is used, the reader is forced to process the words while also trying to develop a mental model of the concept or activity. This creates an overload in working memory and compromises ability to learn (pp. 54–55). By facilitating use of both channels, people can better process information than they can when too much of one system is used.

In a meta-analysis, Clark and Feldon (2005) challenged a number of commonly held beliefs about the effectiveness of multimedia learning. They found that multimedia approaches may not work on all people because of individual differences in learning styles and the nature of information being presented (p. 98). Also, multimedia instruction may not be as motivating a learning tool as proposed. An important consideration in Mayer's (2001) theory is that he accounts for individual differences based on learners' experiences and knowledge prior to receiving instruction. These individual differences limit the ability of scholarship to generalize the benefits of multimedia approaches to learning.

Kalyuga (2005) conducted another meta-analysis of research concerning the "prior knowledge principle" and its impact on multimedia learning. She examined eight particular studies, all of which used experimental designs to measure learning with multimedia learning tools. She found that, across the studies she reviewed, learners who had some prior experience with a similar task were able to learn a new task more quickly than those who had no prior experience with that task (p. 325). This suggests that a trainer or teacher should come to understand a learner's prior experiences with and knowledge about a given task so as to customize the mode of instruction accordingly (p. 334). Such customization may improve learning.

Further, as Mayer (2001) did, Gee (2003) connected prior experience and knowledge to learning. People learn by making connections between past experiences and new experiences (pp. 75–76). Games also, however, teach critical thinking by encouraging users to evaluate their experiences and apply that learning to new encounters (p. 92).

Much as Moreno and Mayer (2000) observed that different people may respond differently to training because of their background experiences, Lacey and Sathian (2012) observe that, "individual history (visual experience, training, etc.)" can affect how each person responds to various sensory experiences (p. 184). Indeed, they call for further study into the relationship of one's background to learning relative to neuron activity. Other neuroscientists have also noted the role of experience in understanding how information is processed (King, Doubell, & Skaliora, 2004; Wallace, 2004).

CONCLUSION

There is considerable overlap in understanding what affects cognition relative to the various scientific fields involved in cognition—social, natural, and physical. As Jack (2012) observes with her introduction of the concept of "neuro-rhetoric," the field of rhetoric can contribute to studies in cognitive neuroscience. The information I have provided here, showing these overlaps, helps to introduce a model that can facilitate interdisciplinary theorization and study specific to multimodal rhetoric and cognition. In the next chapters, I detail the model that can be used to integrate the two disciplines.

CHAPTER 3

The Neurocognitive Model of Multimodal Rhetoric

In this chapter, I identify specific attributes of a neurocognitive model of multimodal rhetoric. Considering the review of extant literature in both fields—neurobiology/physiology and multimodal rhetoric—provided in the previous chapter, five particular characteristics of neurobiology, distributed cognition, and multimodal rhetoric emerge, each of which integrates elements of existing theory, however, I synthesize perspectives from both fields. I also introduce the nature of the studies that act as cases that I use in subsequent chapters to illustrate the model.

A debate in the field of multimodal rhetoric is the orientation of a given theory. As mentioned in Chapter 1, rhetoric includes the way information is presented within a message (design of a message) and how one perceives that message (audience's perception of the information in the message). Kress' (2009) terminology, for example, tends to focus on design attributes of multimodality—items that affect how one is able to develop a given multimodal product and how one considers that product via social semiotics. Others like Ball (2006) as well as Odell and Katz (2009) try to theorize more from the audience's perspective, that is, how to develop a given message relative to how one perceives information and their expectations relative to information needs. Much like the process of usability testing in technical communication and information architecture, I (Remley 2012) attempt to use both approaches as a means to develop good products through an interactive and iterative process—a multimodal object is developed and tested, an audience provides feedback regarding their perception of that object and its ability to help them understand its meaning toward facilitating revision to improve the rhetorical effectiveness, that is, to improve the ability of the message to provide the intended meaning so the audience can understand it better.

Because this theory integrates biological and social attributes, it is necessarily audience oriented, however, it attempts to integrate attributes of design. As is generally considered within the field of technical communication, one must understand one's audience and its information needs before designing a given message. So this model can be applied relative to analyzing a given product and an audience's reaction to its effectiveness as well as relative to designing an effective product to facilitate cognition.

The five particular attributes from a synthesis of scholarship in multimodal rhetoric and neurobiology that emerge are

1. Intermodal sensory redundancy—the preference to integrate more than one sense to facilitate reinforcement of information;
2. Visual dominance—recognition that the visual sense is usually engaged first and as such is dominant in information processing;
3. Temporal synchronicity—the timing of exposure to different stimuli affects how related information is processed;
4. Prior experience—the role that previous experience with certain information or learning style has on acquiring new information; and
5. Attention-modal filtering—that one must filter certain modal information in an effort to concentrate and best process relevant multimodal information.

Any multimodal message or multisensory experience integrates all of these attributes toward affecting perception of information. While it appears to involve more attributes of neurobiology than social science, certain attributes of social semiotic theory are captured throughout this model. The media through which social interactions are facilitated contribute to developing neural structures that support (or not) all of the attributes in this model. Mitchell (1995) observed that we are a visual culture; it is not a coincidence when one considers the prominence of visual stimuli in various media available since 1990: television, the Internet, conference calls. Even radio has entered the visual field with some shows being broadcast on television.

Research can focus on any one or combination of two or more of these attributes. Scholarship in visual rhetoric, for example, has focused on elements of visual dominance while emphasizing social semiotic theories—the impact that our relationships have on our understanding and processing of visual information. However, with this integrated model, it can also consider inclusion of the Colavita visual dominance effect—a biological phenomenon. However, I develop the discussion of research approaches in another chapter. In this chapter, I want to detail each of the model's attributes. First, I need to clarify some terms I use within this discussion. Because I am integrating discourse from different fields, some terms from each need to be explained and reconciled.

TERMINOLOGY

Certain terminology is associated with the attributes that I mention. Some of these are consistent across the disciplinary discourses related to the model. The principle terms that require some explanation are

1. Modal or mode
2. Sensory or sense
3. Filtering
4. Processing

Generally, they are used consistently between the discourse communities involved, however, I briefly explain each to eliminate any potential confusion.

Modal/Modes: As defined by the New London Group (1996), a mode is a particular system of representation. It is the particular means through which particular neurons are stimulated. That is, visual modes stimulate visual neurons of either unimodal or multimodal types. Modal is an adjective that pertains to a given mode or representation system that acts as a stimulus.

Sensory/Sense: Neurophysiological activity is generally described relative to cortices associated with the senses that are stimulated by particular modes of representation. A sense is the biological experience associated with a given stimulus. For example, visual modes stimulate neurons associated with the visual sense. Sensory is an adjective related to sense. Visual sensory experience, for example, pertains to the experience associated with the visual sense.

Filtering: To filter means to distinguish between relevant and irrelevant information. As one experiences particular information, he or she must consider whether that information is relevant to a given message. If certain information is perceived as relevant, the brain concentrates on that information so it becomes part of the cognitive process. Information perceived as irrelevant is removed from further consideration within a cognitive process. However, that act of removing the information is part of a cognitive process.

Processing: Because neural dynamics are associated with the model, the term "processing" pertains to that activity. That is, it pertains directly to the behaviors of neurons toward facilitating cognition. I use the term "cognitive processing" several times. This term pertains to the behaviors of neurons as they facilitate cognition based on responses to stimuli.

Again, most of the terms are used consistently in scholarship from the different disciplines involved, however, this information reinforces their definitions within this theory. I explain each attribute of the model in the following sections of this chapter.

INTERMODAL SENSORY REDUNDANCY

Moreno and Mayer (2000) theorized a redundancy principle relative to integrating similar information in two modal forms in a single message to reinforce the information. The same kind of principle applies to the theory of multimodal neuron processing. Bizley and King (2012) describe the visual-auditory link and how auditory information helps to process visual information, and both can be facilitated via multimodal neurons. Newell (2004) echoed this finding. A basic principle of cognition is the reinforcement of information. This model recognizes this reinforcement with this principle.

Generally, it appears that neurons associated with the first sense engaged attempt to process information toward cognition, however, they want more senses involved. This may be why multimodal neurons exist and become active with sensory information from various modes of representation. Multimodal neurons facilitate processing multimodal information, which, as Arnheim (1969) explained, facilitates a Gestalt effect—a more complete picture of the information. Further, many neurons seem to converge in a portion of the midbrain (Clemo, Keniston, & Meredith, 2012; King & Calvert, 2001). Neurons of all types—unimodal and multimodal—intersect or come to a point where they seem to intersect (Clemo et al., 2012, p. 5). While different cortices are associated with particular stimuli, the neurons all follow paths into the midbrain. Clemo et al. (2012) report on studies of cats and monkeys, but they generalize these findings to humans.

The initial neurons engaged processes information, but other modal neurons reinforce or further define information, much as theorized by Moreno and Mayer (2000). For example, visual information shows what an object/abstract concept is or looks like. Other modes contribute to refining the definition or composition of the object. In the example provided by Moreno and Mayer—learning about lightning—an image of a house, clouds, and lightning are provided, and text labels various attributes of the process involved in generating lightning. The pictures of the objects facilitate some information, but without the text, it would be difficult to comprehend the process fully. The study group related to the animation and narration combination (visual-audio) was better able to learn information than was the group that had animation and text (strictly visual).

It also reinforces what Mitchell (1996) stated about ekphrasis, the concept related to optimal combinations of modes toward best articulating a message. Information related to a single sense can facilitate cognition, however, information from various modes facilitates cognition better. Perrault, Rowland, and Stein (2012) call this "multisensory enhancement" or multisensory synergy. Connecting neurophysiology to behavior, they observe that multisensory inputs tend to "elicit more vigorous responses than are evoked by the strongest of them individually" (p. 281).

This principle is echoed throughout the cases presented in subsequent chapters. For example, Chapter 5 discusses it relative to simulators and how they engage visual as well as tactile and spatial senses. Flight simulators, specifically, place a student pilot in the environment in which he or she would operate and allows that student to experience all the sensory stimuli a pilot flying a particular aircraft would experience.

VISUAL DOMINANCE

There is a plethora of empirical data related to visual processing and visual rhetoric. An entire subfield within both areas of rhetoric and neurobiology works with this scholarship. Much of it pertains to relationships between visual information and combinations with other senses. Arnheim (1969) and Mitchell (1996) emphasized the visual attributes of cognition, even in the titles of their respective texts—*Visual Thinking* and *Picture Theory*. While most of this attribute comes from neurobiology, some dynamics are recognized in both neurobiology and social semiotic theory.

Again, neurobiology scholarship finds an attention-related preference for visual information. If visual information exists along with other stimuli, then the visual will draw attention no matter what other stimuli are involved. Generally, the first sense engaged has priority in processing information. However, the visual is perceived before any other stimulus, which affects perception of effects of various modes. According to Spence et al. (2012), studies related to varying the timing of exposure to audio and visual stimuli and awareness of one being perceived before the other find that only when audio stimuli are presented at least 600 milliseconds before the visual stimulus is presented do participants perceive the audio to precede the visual. When multiple stimuli are involved at approximately the same time and vision is one of them, the visual will be attended to more than any other stimulus. Spence et al. note that scholarship on the Colavita visual dominance effect finds that even if one is directed to attend to audio stimuli when the two—visual and audio—are provided at the same time, one attends primarily to the visual (p. 534). They also note that if the presentation of audio stimuli precedes visual stimuli, that can affect the degree to which one attends to visual stimuli (p. 534). Regardless, they state that "although attentional manipulations can sometimes be used to modulate, or even eliminate, the Colavita visual dominance effect, they cannot be used to reverse it" (p. 534). Once the visual sense is integrated, attention to that stimulus will persist.

In the absence of direct visual stimuli, one tends to imagine or look for an equivalent based on engaged senses. As Mayer (2001) observed, one may "attempt to build a visual model" (p. 63). When one is talking to someone else on the telephone, he may envision the person with whom he is speaking if they are acquainted. If listening to a radio broadcast, one may try to recall a picture of the radio personality who is talking or a photograph of the band if listening to music. This reinforces the modal redundancy principle: One wants as much stimuli to help process information. If the visual stimulus is absent, the brain will attempt to create a visual stimulus to assist cognition. One remembers what the person looks like from previous interactions and experiences with them and envisions them.

Visual multimodal neurons dominate because more information is contained in visual information than other senses. There is an appeal about the Gestalt effect, identified by Arnheim (1969). One can assess spatial relations, textures, size, distance, proportions, and image within just visual information. While other senses may provide refinement of the assessment related to those items, multiple types of visual information are provided in one image.

Social science research has theorized how interactions and technologies through which they occur affect perception and cognition. I discuss the influence that technology has in the next chapter, however, I describe relevant theories associated with the visual dominance attribute of the model here. Media technologies that facilitate visual information/visual forms of information encourage the visual dominance effect, also facilitating "visual culture." For example, two particular theories of communication technology are ideal for integrating social semiotics into the discussion. These are media richness theory and social presence theory.

Media richness theory argues that given media will facilitate effective communication for a given task. Different media are required for different tasks because of the level of richness or ability to minimize uncertainty or ambiguity (Bouwman, van den Hoof, van de Wijngaert, & Dijk, 2005; Rice, 1993). Traditional face-to-face communication is considered the richest medium. The listener is able to see the communicator's visual cues and hear voice inflections that help to understand a message, while the speaker is able to see the listener's reaction and respond to any potential confusion immediately. Potential for confusion is limited because of the multiple stimuli available to assist in cognition.

Similarly, social presence theory considers the degree to which communicators feel the other is "present" while communicating. Generally, it is perceived that more "social" communication can occur when communicators are physically present to each other, as in face-to-face communication, while more formal communication is likely to occur when that physical presence is lacking, as in e-mail (Hewett, Remley, Zemliansky, & DiPardo, 2010). Social presence theory attempts to understand to what degree certain media facilitate social presence, and therefore more effective communication. In both theories, the ability to see and hear one conveying a message is considered the best to facilitate cognition.

Video tutorials are a popular form of instructional tool for asynchronous learning. Such videos allow a viewer to see the narrator or the task being taught in the absence of a real face-to-face interaction with the teacher or trainer. Detailed case studies that include such a video product and the application of this principle are provided in Chapters 6 through 8.

Related to this intersection of visual rhetoric, technology, and cognition is that medical scans contribute to cognition about a condition and shape medical decisions. Teston (2012) observes within deliberations about cancer patient care that "the visual makes possible the construction of knowledge about disease" (p. 205). Even within the medical field, decisions are being made based on visual information, reinforcing the powerful role played by visual stimuli and technologies that closely reproduce the material.

Again, scholarship in visual rhetoric and neurobiology recognize the preference and esteem associated with visual modes and sense. That neurobiology recognizes biological attributes that account for it can help advance our understanding of multimodal rhetoric.

TEMPORAL SYNCHRONICITY

Per studies by Keetels and Vrooman (2012) as well as Moreno and Mayer (2000), sensory information is processed at different rates. Neurons are able to process sensory information at different rates relative to the modes involved. While the Colavita visual dominance effect recognizes this, it emphasizes any combinations that include visual modes. The principle of temporal synchronicity considers combinations in any modal form, including those that use the same mode, such as two forms of visual information, print-linguistic text and video, for example. Such an issue is important to a theory of cognition because the rate at which information can be processed affects the rate at which cognition may occur. Multimodal rhetoric involves ascertaining optimal combinations of modes to facilitate cognition.

Moreno and Mayer (2000) had four different modal combinations relative to timing of the information. In their studies, they found that when information from visual and auditory modes was processed at the same time, learners were able to acquire information better than when the same kind of information was facilitated via two kinds of visual information at the same time (animation and text). However, they also found that learners tended to transfer knowledge better when they learned via a sequence of different modal information—watching an animation and then hearing a narration about what they viewed.

Such findings indicate that the types of modes involved and timing of information provided affect the efficiency of processing it. As mentioned above, when audio stimuli are presented several milliseconds before the visual stimulus is presented, participants perceive that the audio precedes the visual. Consequently, timing of exposure associated with certain modal combinations becomes part of the rhetorical effectiveness of those combinations. As considered with the Colavita visual dominance effect, the visual is the fastest sense engaged, but it takes the longest to process information. However, sound takes less time to process. When both visual and auditory senses are stimulated, both are attended to by the audience. Perhaps this influences the first attribute stated above: intermodal redundancy and the desire for more senses to be engaged in cognition. That is, combining senses to help facilitate cognition enables the system to process information faster. As indicated earlier, though, the timing of sensory engagement or modal presentation can affect the degree to which particular neurons are engaged and consequently how information is processed (Spence et al., 2012).

An example of this attribute detailed in an extended case study in Chapter 6 shows that combining information may lead to an information overload. In a video product developed by a company that supports soccer coaching and training, visual information is presented on the screen in multiple forms—animation and print-linguistic text. However, these are separated such that one is shown just before the other, allowing the viewer to process information better than if showing them at the same time.

Just as Moreno and Mayer (2000) found that certain combinations of graphic and audio information and the timing related to those combinations affect learning, the issue of temporal synchronicity comes into play relative to other combinations as well as biological attributes of the audience that can affect processing. For example, immersive learning environments like internships or virtual environments/simulations engage all or almost all of the senses relative to various related modes included—visual, spatial, audio, and possibly smell and touch. These particular dynamics will be discussed more fully within a detailed analysis in a later chapter. There is scholarship in both neurobiology and multimodal rhetoric regarding new media such as virtual learning as well as real training. I detail relevant literature pertaining to simulations and hands-on training processes and how they affect cognition neurologically later in this book.

Also, head injuries and brain disorders that affect the various cortices impact how the brain can process information. Consequently, scholarship in neurobiology/physiology is considering more precisely how such injuries and disorders affect what may be considered effective rhetorical combinations within learning and perception generally. Examples of such research include autism, brain injuries, lost or low vision, deafness, and neurological disorder or injury that causes numbing of touch-sensory experiences. As mentioned in the previous chapter, neurophysiological scholarship finds that combinations of visual and touch senses contribute to understanding how to interact with physical objects: If one can hold an object but not feel it, how would that impact one's perception of it?

PRIOR EXPERIENCE

Scholarship associated with the general field of cognition observes that prior experience plays a major role in learning. As mentioned in Chapter 2, prominent works by Pinker (1997), Gee (2003), Moreno and Mayer (2000), and Hutchins (1995) echo this understanding about learning. Unless affected by memory disorders, one cannot help but learn from experience. As one experiences a given situation, he or she recalls a similar experience and consequences associated with their actions related to it. If he did not like the outcome of his actions, he reflects on what he could have done differently and changes his behavior accordingly to try to get a better outcome the next time it occurs. With each experience, one changes their actions to try to arrive at a favorable outcome related to the experience. This is the process of learning to behave a certain way to bring about a given outcome relative to a particular context or situation. These situations, though, tend to be social, so this attribute involves much social science. However, it also influences neurons, because neurons grow synapses with each social experience that reinforces certain information or knowledge. The more synapses one has developed from previous experiences and learning, the faster one can process related information.

Spence et al. (2012) as well as Lacey and Sathian (2012) also study how prior experience affects neurophysiological processes. They find generally that experience contributes to changing neurological behaviors (plasticity). Plasticity is a lifelong process: Different kinds of changes occur at different periods of one's life. Hoiland (2012) notes that neurons grow in developmental years and some die out as we age. Gopnik, Meltzoff, and Kuhl (1999) asserted that there are approximately 2,500 synapses per neuron at birth; this grows to approximately 15,000 synapses per neuron by the time one is 3 years old. However, the average adult has only about 7,500 synapses per neuron. Experience dictates which synapses are deleted or "pruned" (Gopnik et al., 1999). Generally, as one learns a given task, more synaptic links of certain types develop, facilitating faster processing of information. Clemo et al. (2012) report that in their studies of monkeys and cats, the average number of synaptic connections for a single neuron is two to three, but some may have as many as four connections (p. 11). As one learns from experience, more connections are developed for each related synapse, and these connections help to process information faster.

Such changes to neurons, stimulated by patterns of learning approaches, contribute to affecting perception and attention attributes of information processing. Consequently, this attribute can involve analyses of plasticity and synaptic connections. Such dynamics may also affect whether one is visual learner or less so. What instructional/learning methods has a person been exposed to earlier in life/development? Is one able to process information conveyed via radio transmission as readily as one experiences a television broadcast? One who has a lot of experience watching sporting events on television but has rarely experienced a radio broadcast of a sports event, for example, may not be able to follow a radio broadcast. Radio broadcast of sports events involves imagining action based on description; television facilitates both visual and audio information.

As suggested with the references to Gee (2003) and Mitchell (1995), learning via social interactions affects what is learned and how. This is another area in which social science theories contribute to the integrated theory. As Hutchins (2000) noted, distributed cognition involves any system, whether it be neural systems or systems of groups of people (p. 2). Hutchins also acknowledged that it is through interactions with artifacts as well as with others that cognitive development occurs (p. 5). Such interactions affect neural plasticity within the individual's brain; social interaction is a part of that neural development. It is generally recognized that learning occurs through experience.

Culture is a social phenomenon, so visual culture is a social valuing of visual information. However, social elements of technology also are involved in the use of technologies—visual/audio/other—including simulators. Gee (2003) and Mitchell (1995) separately noted the increased use of simulators for training purposes, and a subfield of scholarship in cognition examines such uses of simulators and related technologies. I (Remley, 2009) also noted the use of hands-on training to help people transition to new kinds of work using a training approach similar to what they received previously. Such scholarship finds that when people learn how to do particular tasks a certain way and that way has been used throughout their lives for learning new tasks, people are more likely to favor that approach, be it visual, hands-on, or otherwise.

Relative to learning and cognition, one may learn better with modes he has experience using to learn new concepts. If one, for example, is accustomed to learning new tasks through hands-on training, then she may not learn well with only audio and textual instruction. Also, the information that one possesses, based on experiences, affects their processing of new information or their perception of certain information available to them. The example of the carbon monoxide poisoning situation that I mentioned in Chapter 1 illustrates this point.

In that example, I mentioned that the fireman who responded to my non-emergency phone call about the new fireplace immediately commented about the color of the flame and that such a color indicated a problem. I never had experience with such information—flame color and its relationship to carbon monoxide dynamics—so I was unable to make that link. The fireman and I experienced the exact same visual information, but the knowledge he possessed through previous training helped him to understand that relationship. Consequently, he had previous knowledge that helped him process relevant information so as to understand the situation much better than I did. Further, through that social experience, I have learned how to process that information better so as to understand such a problem. If I observe a similar color within a gas flame, I will know it indicates a problem.

Another example within a larger case study is in Chapter 5, associated with using the prior experiences of how a group of people tended to learn to help them learn new work. Within that historical study, government and industry combined to develop a training model that uses a form of hands-on training to help workers learn war-industry work. This hands-on training method is akin to how the same group learned how to perform farm-industry work, work from which they came.

Our prior experiences contribute to shaping our understanding and ways we come to understand new concepts. This principle recognizes the role of that experience in cognition.

ATTENTION-MODAL FILTERING

The last attribute of the integrated model pertains to the literature about filtering of information and attention. Moreno and Mayer (2000) described the attention principle related to learning with multimodal tools. When irrelevant information is included in the information, it negatively affects cognition: The brain must process too much information, and it has to filter what it perceives to be relevant and what it perceives to be irrelevant. The more modes there are involved, the more challenging this is. Further, Moreno and Mayer recognized the Baddeley (1986) principle of short-term memory and its ability to process information. That is, the more modes there are involved, the more processes are involved, and that negatively affects the ability to process information. Baddeley as well as Moreno and Mayer encouraged limiting modes to two.

Tufte (1983) also observed that irrelevant information negatively affects the ability to understand a given message. He termed such information "chart junk." Generally, it is textual or visual information that is not required within a given image or graphic in order to understand the meaning of that graphic. Its inclusion gives the initial impression that it is relevant, and the brain attempts to process it. Consequently, the filtering process related to ascertaining which information is relevant and which is not slows down cognition. This effect on the brain and cognition is observed in neurobiological/physiological scholarship.

Campos and Bulthoff (2012) find this relationship relative to virtual environments: The less irrelevant information that is contained in a message, the better one is able to process it quickly. Irrelevant stimuli must be filtered to get to relevant information. So, reinforcing scholarship in cognitive psychology and education, the more one can eliminate irrelevant information, the faster one can process the information, contributing to better learning.

This principle is illustrated in Chapter 7 relative to a student orientation video. The student who created the video included too much irrelevant space in a particular scene. A viewer, consequently, could become distracted or have to filter that space out.

MEDIUM

The principles described so far are part of the design of the message relative to how it facilitates neural processes toward cognition. The rhetorical situation includes consideration of the audience, purpose, and medium or media used to convey the message. Neural attributes are part of the consideration of the audience, and the purpose within the parameters of this book is instruction. The model considers the medium of delivery as a framing attribute. In multimodal rhetoric, the medium engages multiple senses. Cognition involves multiple senses, and various media engage multiple senses through the modes of representation they use. Consequently, the medium or media used for delivering the message influences what stimuli can be used within the instructional message. A given medium facilitates or restricts the stimuli used for instruction, affecting neural processing too. Consequently, part of the discussion I present reflects a critique of the technological tools associated with the product used in the case study chapters.

I use cases associated with a number of different media—face-to-face interaction, multiple forms of video, simulation, and slide shows. The technology associated with the media shapes the message, and one needs to consider the affordances and constraints of a given tool to develop an effective message. I detail this consideration in Chapter 4.

Figure 3.1 represents this model visually. Each attribute is represented, though, not in any particular ordering. The model integrates elements of the multimodal product/composition, but it focuses attention on the product's ability to engage an audience's neurological experiences and perceptions of information toward cognition. Different combinations of modes provided in a given piece, affected by the technology available with the media, are the objects of study relative to all but "prior experience." Intermodal sensory redundancy examines the particular combinations at work and their resulting sensory experiences. Visual dominance suggests the degree to which the visual mode is dominant in a given piece, prompting the visual sensory experience. Temporal synchronicity studies the timing of presentation and how that timing affects perception and processes. Modal-attention filtering examines the degree to which one distinguishes between

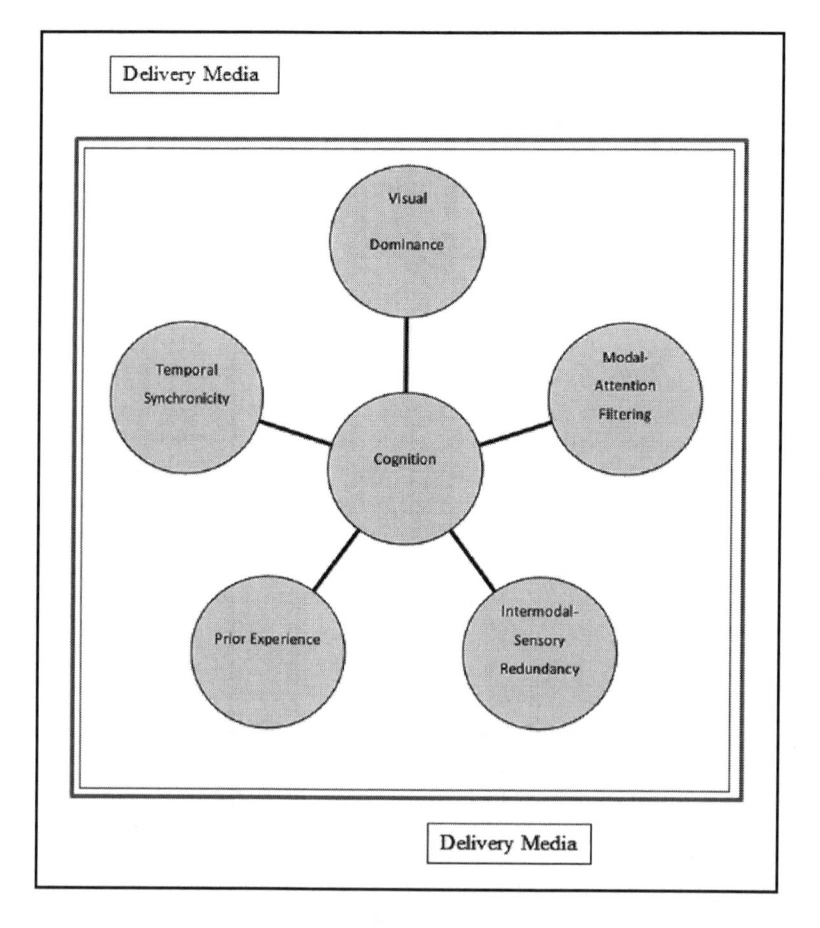

Figure 3.1 Model.

relevant and irrelevant information and filters out what he or she perceives to be irrelevant and why.

Each attribute is represented independently because each can be studied as a unit of analysis in itself. That is, each may be isolated within an analysis while others are maintained relative to the product or audience, however, it is important to understand that all contribute to cognitive processes. Further, because various dynamics occur within each attribute, one may study a set of dynamics, much as social semiotic scholarship has tended to focus on the social attributes of, for example, prior experience. So existing scholarship can still fit into this model.

Surrounding the attributes described above is the medium used for delivery, which frames the message and shapes the means by which cognition can occur. I detail the frame in the next chapter. Each attribute, then, represents an attribute of the message that engages neurological and semiotic dynamics that facilitate cognition.

STUDYING MULTIMODAL RHETORIC THROUGH THE SOCIONEUROBIOLOGICAL LENS

As mentioned in previous chapters, the fields of rhetoric and neurobiology/physiology tend to examine multimodal and multisensory experiences differently, using different tools especially. While rhetoric studies tend to focus attention on composition and observed behaviors or surveys of audience perception of content, neurophysiological studies focus on particular neuron behaviors inside the brain based on biomedical technologies. However, these neurobiological/physiological studies involve analysis of such activity relative to certain stimuli; nevertheless, social science is involved in neurobiology/physiology. As such, this model can also be applied to rhetorical analyses of multimodal products/stimuli to form a neurorhetorical analysis. Consequently, rhetorical theory, often linked with social sciences, can contribute to cognitive neuroscience studies, and the field of rhetoric can benefit from integrated studies likewise.

Examples of potential studies include ascertaining how the brain processes certain modal combinations; rhetoric scholars can design these combinations. Studies can triangulate data by including both biomedical technologies as well as social science research methods, such as surveys and interviews with participants as well as observation and quasi-experimental designs. I discuss such studies in later chapters. However, the chapters immediately following this one detail neurorhetorical analyses of specific products without triangulation.

Theoretical Limitations

A limitation of this model is that it assumes a reasonably normal neurobiology/physiology is at work, however, it also facilitates theory development relative to neural disorders or the absence of certain neurophysiological elements. As mentioned above, numerous studies in neurophysiology consider the biology associated with low vision or blindness and relative to other disorders. Such disorders suggest damage to the cognitive system referred to in Chapter 1. The brain compensates for what it lacks to facilitate cognition, however, such disorders present challenges to understanding how neural processes work in damaged systems.

An example of such a limitation on the application of this model is the case of Phineas Gage, generally considered "neuroscience's most-celebrated case." Phineas Gage was a railroad foreman in the 1840s who was generally considered a good worker and responsible foreman. One day, as he was tamping gunpowder into a hole, preparing the ground for a blast to create an opening for more rail, the powder ignited. This explosion sent the tamping rod through his left cheek, exiting through the top of Gage's head. A hole of about 1½ inches was left at an angle through the left side of his head. Remarkably, Gage was not killed in the accident, and he was even able to go to a doctor for treatment. However, all documentation indicates that Gage's behavior changed dramatically after the accident: He was no longer able to control his emotions and treated others with considerable disrespect. Studies of the case find that, in addition to injury to the neurons associated with cognition, white matter was severely damaged. White matter of the brain is fatty tissue that facilitates transmission of neural activity (Van Horn et al., 2012). His cognitive system was damaged in multiple ways affecting his ability to compensate.

Such cases notwithstanding, I provide a brief example of case study analysis integrating this theory in the next section. I use the anecdote I provided in the first chapter, specifically that of my experience with the fireplace. The experience involved multiple senses being engaged at different times, which affected my ability to process relevant information to understand what was happening. It also involved learning about relevant information associated with a particular sense that led to a change in my ability to perceive information.

Another limitation is that it is not comprehensive, as I stated in the Preface. It is limited by my own scholarship and background. I expect the model to grow to include additional principles as other disciplines interact with and use it.

A BASIC EXAMPLE OF ANALYSIS

To review the basics of the particular anecdote, I had purchased a gas fireplace and as the installers finished their job, they indicated there would be some odor as paint burned off, and the odor would be gone after a few hours of burning. I allowed the fireplace to burn for a few hours that same evening. The next day the odor continued, and I also had some dryness in my mouth and felt light-headed. Concerned about the dry mouth and lightheadedness, I contacted the fire department's nonemergency phone number and asked someone to check on the fireplace for anything that may be a concern, and they wound up sending a fire engine as if responding to an emergency.

I begin my analysis relative to the senses engaged to raise concern about the fireplace. I could see the fireplace and the flames associated with it. I could hear the burner as it burned gas. I could sense the heat from the fireplace (touch). I could also smell the odor I was told about. I include lightheadedness and dryness of mouth as haptic (touch-related) sensory experiences. While visual stimuli were present, I did not perceive a problem from that information. However, from the smell and touch senses, I perceived a problem. Prior information and the timing connected to my experiencing them helped me understand that what I was sensing could be problematic.

I was aware from previous literature about fainting and concussions and my own experiences with concussions that lightheadedness, generally, signals an unnatural state. The information I already had suggested a problem associated with that sense. I connected my lightheadedness to the fireplace since I had not experienced such a condition prior to its installation. I also associated it with the fireplace because of the previous information conveyed by the installers. I was told that the odor should dissipate in a few hours of burning, and that did not occur, so I perceived that to signal a problem generally. I was experiencing multiple sensory processes at the same time, and at a particular time (shortly after the fireplace was installed), that suggested something was wrong.

However, interestingly, even though I could see the fire and related colors in the fire, I did not perceive a problem from that information. I had no prior experience with information related to those dynamics, so I was not able to process that particular information toward understanding a problem. Indeed, because of my lack of knowledge about its relationship to what I was experiencing, I filtered it out and did not consider it within my cognition. Each time I looked at the fireplace, I could see the flames and the color, which included colors found in a regular fire—yellow and orange as well as blue. The literature I had pertaining to the fireplace indicated that it would be a natural-looking flame, so I perceived the colors to be normal given that information. I had a visual experience, however, I misunderstood the information related to it because of a lack of knowledge about it within the particular context.

So, while I did not know of a particular problem with the fireplace, my own perception of a problem at all was influenced by the various multiple stimuli available to me and my own prior experiences related to those stimuli. My prior experiences (and lack thereof in terms of the visual) also influenced which modes I paid most attention to as I processed the total information available to me.

My cognitive process related to perceiving a problem with the fireplace integrated all attributes listed with the theory. However, the Colavita visual dominance effect actually had a negative impact on my perception because of lack of information related to that information. My perception changed with the visit from the fireman.

While one could conduct an analysis related to the cognitive process related to my perception of the fire engine relative to the siren and the fire department's perception of this as an emergency, I will permit someone else to do those. I will focus attention on the fireman's visit and how it affected my own perception. After questioning why a fire engine was sent in response to a nonemergency call, I invited the fireman into my house. His reaction and observation changed my ability to understand the visual information associated with the fire colors.

Firemen receive considerable training toward understanding various attributes of their job and fire emergencies. Among this training, evidently, is learning how to distinguish colors within a fire and what different color combinations indicate about oxygen content within the fire. The fireplace was positioned within view upon entrance to my house from the front door, where the fireman entered. His initial reaction was relative to the visual information he saw in the color of the fire, which certainly suggests the Colavita visual dominance effect.

Indeed, understanding that it was a gas fireplace and seeing inappropriate colors in the flames, he immediately stated that the fire should be bluer than it was. I asked further about what he saw in the flames. He stated that it looked like the burner was not processing oxygen in the air well such that the oxygen content in the fire was poor. He encouraged me to get the CO detector, consequently.

With this information, I researched color composition of gas fires and found additional information reinforcing his conclusions. Information in the owner's manual that came with the fireplace did not refer to any of this information, so it did not facilitate such cognition as well as the materials I found in that research. One would be able to compare the different materials to ascertain, which facilitated certain information more effectively relative to a given purpose. However, the manufacturer may not have attempted to facilitate that understanding with the materials it provided. The information they provided may be considered appropriate for marketing purposes. The information I needed, though, would be appropriate for a troubleshooting section within a manual.

The CO detector also reinforced his conclusion about the problem with the burner toward diagnosing it specifically and quantitatively. Not only was there a problem with the burner, but it qualified as an emergency situation. My understanding of this color composition has changed the way I look at flames in a gas fireplace. Consequently, there is some degree of plasticity associated with this experience that will enhance the Colavita effect on me in the future relative to similar situations. Because I have more information based on this information and experience, my perception of the related visual information has changed.

CONCLUSION

The information in this chapter identifies and describes the particular attributes characteristic of cognition involved with neural processing of multimodal stimuli. As such, it provides a foundation by which to analyze audience responses to multimodal information toward developing a better understanding of multimodal rhetoric. Also, I provided a brief application of the model to an anecdote.

The next four chapters detail analyses of multimodal rhetoric associated with learning relative to this model to illustrate its applications. A few of the cases are derived from data I collected for previous studies not directly related to this theorization, while two chapters include new data that I analyze within the model's structure. While the analyses, like the model, draw upon an understanding of neurobiological studies, the discussion of the application is theoretical rather than measured with specific empirical tools. However, in a subsequent chapter, I discuss how empirical studies can be developed toward further application and development of this model.

CHAPTER 4

Framing Cognition with Media

INTRODUCTION

Different media facilitate different modes of representation, and a medium brings with it different attributes to consider in assessing rhetoric relative to the available modes (Sorapure, 2005). This point also impacts a given medium's ability to facilitate learning relative to those modes. A video, for example, facilitates more animation than a slide can within a slide show, however, more digital space is needed to access and view a video than to view a slide show. Further, different slide show tools have different capabilities.

The tools available to both composer and audience become part of the multimodal rhetorical situation, and this impacts the model of cognition as well. Much as an audience's biological attributes affect their ability to learn, the technology used to design the instructional materials affects which modes of representation and stimuli are included and how they are included. If I have access to only pen and paper to convey a message, that limits the design of the message considerably more than if I have access to a word processor like Word and a graphics tool like Photoshop to compose the message. The medium or media used to facilitate learning is included in the model's principles as a framing attribute, but it does not favor a single medium over others generally. I address this attribute of the model in this chapter by offering suggestions to enable the tool's capabilities to become part of the model's dynamic. Assessment theory helps facilitate this consideration.

Scholarship in Assessing Multimodal Projects

A growing body of scholarship theorizes ways to evaluate and assess multimodal products in writing and technical communication courses. For example, a 2012 issue of *Technical Communication Quarterly* (Odell & Katz, 2012) provides some insight as do articles in *Computers and Composition Online* (e.g., Murray, Sheets, & Williams, 2010; Remley, 2010, 2012). I call attention to some of the points presented in this scholarship about assessing multimodal projects, because it contributes to understanding rhetorical effectiveness of such compositions toward facilitating learning and cognition; but I do not theorize such assessment here.

As I indicated in the first chapter, rhetoric is about understanding how to present information to encourage a certain kind of response from the audience to that information relative to a given purpose. Scholarship providing rubrics related to assessment of multimodal products struggles to define or characterize combinations that could be used to ascertain what effective multimodal rhetoric is. However, studies such as Moreno and Mayer's (2000) help ascertain attributes of effective multimodal rhetoric within a given context and purpose. While some scholars attempt to use design-oriented attributes of rhetoric for assessment, others encourage attributes associated more closely with the reading experience or the learner's cognitive experience.

Ball (2006) observed that many rubrics that facilitate assessment of new media projects tend to emphasize attributes of designer skills; she called for assessment to focus more attention on the rhetoric involved—taking the reader's perspective (p. 394). Odell and Katz (2010) also encouraged an audience-perspective approach. However, Neal (2010) observed that, while writing pedagogies are changing with the emergence of various kinds of digital texts and composing technologies, approaches to assessment "currently do little to account for such change," and he articulated the challenge in recent scholarship to develop a new set of criteria for assessment to facilitate "rhetorical and contextual composing" (p. 757).

While certain attributes of rhetoric can apply across modes of representation, as identified by the aforementioned scholars, the capabilities and limitations of the media and related tools used to compose the message affect design decisions and consequently the rhetoric of the message. The tools to which a composer has access to compose a message and to which an audience has access to view it become part of the rhetorical situation. So a question regarding cognition and rhetoric that emerges is, Does this multimodal product apply effective rhetoric associated with cognition relative to the capabilities of the medium and tools used to create it and neural functions of cognition? I respond to this question within the model by using affordances and constraints of the media of delivery as a framing principle.

Analysis of Affordances and Constraints

Norman (1988) explained that one should study the psychology associated with use of a given technology and its capabilities. A user considers what a given tool can or cannot do as he or she attempts to use it for a given purpose. The affordances and constraints of the technology affect how one can use it. According to Norman, affordances are "the perceived and actual properties of the thing . . . that determine just how the thing could possibly be used" (p. 9). A constraint is something that limits the way in which an artifact can be used. While visual cues associated with the tool can suggest the possibilities of use or limits of use, Norman also acknowledged that affordances and constraints may be part of the medium's design but not immediately visible to the user.

I have noted previously that affordances and constraints of a given tool are relative to the user's previous experiences with similar tools or technologies. While Norman (2009) emphasized firsthand experience as he discussed affordances and constraints, secondary research, readings based on others' experiences, can inform an analysis of affordances and constraints. If one has limited experience with a given tool, her understanding of that tool's capabilities and limitations is shaped as she reads about how someone else used that technology. As Norman stated, "Affordances can signal how an object can be moved, what it will support. . . . Affordances suggest a range of possibilities, constraints limit the number of alternatives" (p. 82). A constraint tells us in some way not to use it a certain way or limits how we can use it.

Previous experience is part of the model; this is associated with an audience's previous experiences with a given concept or tool. However, a designer's own experiences with a given technology frame how he or she uses that tool to design the message. Many new technologies facilitate learning how to use them by incorporating attributes of older but similar technologies. As one looks at the interface of the audio recording application Audacity, for example, one experiences the interface of old tape recorders. As one uses a given technology more often, he or she becomes more familiar with its capabilities. That experience can enable the designer to develop better materials with that tool. Until that happens, though, the design will be limited by the designer's understanding of the tool's abilities.

The Rhetoric of Multimodality

In the past 15 years, a focus of study within multimodal theorization has been on various modal combinations and how they affect meaning-making (Gee, 2003; Lemke, 1998, 1999; Mayer, 2005; Moreno & Mayer, 2000; Richards, 2003; Whithaus, 2012). Studies pertaining to this analysis seek to understand rhetorical attributes of mixed modes and when and under what conditions certain combinations are most productive (e.g., Lemke, 1998, 1999; Richards, 2003). Indeed, several theories of multimodality and this relationship between print-linguistic text and image have been presented, but each seems to have its own difficulties meeting the challenges of theory development thereof.

An example of such a study is that of Moreno and Mayer (2000). Generally, Moreno and Mayer found that certain combinations of visuals and text information affect learning, suggesting a relationship between modes used to communicate and their rhetorical impact. Relative to the rhetoric of instruction, combining visual and verbal/aural modes of representation is more powerful for accomplishing the instructional purpose than using only narration or visuals alone. From these experiments they formulated several principles associated with multimedia instruction. Mayer (2001) summarized their multimodal principle with the statement that people learn better when pictures and words are integrated into an instructional message than when only words are used (p. 63). When only words are used, people may attempt to "build a visual model," but they may not attempt to do so. If a picture is provided, people can make the visual connection more readily. Mayer also asserted that it is vital to eliminate extraneous material—words, images, and sounds—from any multimedia message. Such irrelevant information "competes for cognitive resources in working memory," disrupting the learner's ability to organize and retain relevant information (p. 113). Tufte (2003) also made this point relative to PowerPoint slide shows.

Gee (2003) asserted that how people read and think about a particular thing is determined by their experiences with certain social groups. Through social practices, such groups "encourage people to read and think in certain ways, and not others, about certain sorts of texts and things" (p. 2). Learning is a social practice; it occurs in some kind of social setting generally. Gee identified a marriage between the semiotic domain and situated practice (p. 26). Reading and writing print text is an example of a semiotic domain. Combining notions of socially constructed learning and semiotic domains, this also brings to mind the importance of understanding a user's existing literacies when using certain approaches or modes in training. How much time one needs in order to learn the new mode will affect how quickly he or she will learn the task being presented. It is important to understand modes in which trainees have learned previously, and this is included in the model's principles. Experience builds synapses that enable one to understand a concept that is similar to something they have already learned faster than if the concept is not similar to a previously learned topic. However, Vygotsky (1978) also applied this attribute of socially constructed knowledge to the use of tools to complete an activity (activity theory). One learns to use a given tool based on how they observe others use that tool.

Assessment of the Multimodal

As I mentioned in the opening section of the chapter, some scholars have tried to approach assessment from different perspectives—some from the designer's perspective and others from the audience's perspective. For example, Odell and Katz (2012) called attention to the reader's perspective and needs. Morain and Swarts (2012) and I (Remley, 2012) separately identify particular elements of information design that can be applied in video and other media as well. These include what and how much information is visible on a page or screen, the pace at which narration that accompanies any images moves, how information and content are organized and transitioned from one point to the next, and other attributes affecting the ability of the audience to understand a message relative to the message's purpose. Neal (2011) also included elements of pace and the ability to follow the narration as well as effective transitions (p. 96).

In each of these approaches to assessment of multimodal products, there is sensitivity to the audience's needs and what will help the audience respond a certain way to the information. However, a new rubric emerges with almost each scholarly publication on the topic. In reflecting on assessment approaches and rubrics of authors of 10 different chapters in an edited collection, Herrington, Hodgson, and Moran (2009) observed that "assessment criteria were not generic; they were tailored to the nature and goals of each project" (p. 204). I have argued that one of the variables confounding multimodal assessment scholarship is that different media facilitate different modal combinations and designs (Remley, 2010).

The framing principle of the technology's affordances and constraints considers the diversity of potential media and semiotics with each, integrating a criterion like that quoted by Herrington et al.: "Employ the affordances (capabilities) of the medium you're using in effective rhetorical ways" (2009, p. 205). Effective rhetoric depends on the purpose of the message and the targeted audiences, as found with the Moreno and Mayer (2000) studies relative to instruction and learning. Just as the concept of rhetoric generally begins with an understanding of audience and purpose to facilitate designing a message, this model recognizes the media available for designing a message to facilitate cognition for the purpose of learning.

Technology and the Message

Information may be presented one way using a PowerPoint slide and another way using a web page template. Further, a given tool may not allow certain kinds of animation while another can manage dynamic images like animation or movies: Dreamweaver is able to integrate movies and animation more readily than a design tool like Publisher can. Such differences in technological capabilities affect how well a message can be presented in each medium.

Sullivan (2001), for example, observed dramatic differences between web page design and page design from a book. While she suggested that using a book page design may be the safest way to design a web page, she also acknowledged that because of the different ways people read multimodal pieces, there is "no clear arbiter of what constitutes safe visual rhetoric on the Web" because of its dynamic design (p. 117).

Perhaps the strongest statements about multimodal designs come from Moreno and Mayer (2000) and Tufte (1983, 1990, 2006). As mentioned above, Moreno and Mayer identified several principles about multimedia instructional tools and learning. Those associated specifically with the tool(s) and their capabilities include

a. Split-attention (learning occurs when one does not have to split attention between multiple sources of mutually referring information);
b. Modality principle (better learning occurs when verbal information is presented in auditory mode rather than in visual mode on screen);
c. Redundancy principle (if visual information is presented simultaneously with verbal information, students learn better with an animation/narration combo than with an animation, narration, text combo);
d. Spatial congruity principle (better learning occurs when on-screen visual and text materials are integrated rather than presented separately); and
e. Temporal congruity principle (similar to spatial congruity—involves timing, though).

Moreno and Mayer (2000) thus note how the medium impacts the rhetoric and design. Two important principles that Tufte (1990) encourages apply directly to the capabilities of the tool—elimination of "chart junk," or irrelevant information; and proper "data-to-ink" ratio, so that a graphic is not cluttered by too much information. Both of these pertain to use of space, which is affected by the size of screen that facilitates viewing and screen resolution. Rephrasing these attributes relative to the tools' abilities,

a. How well can the tool synthesize a limited number of modes to facilitate a given message?
b. How well can the tool integrate multiple modes simultaneously or asynchronously?
c. How well are two particular modes that facilitate information well in combination integrated using the tool?
d. How well can space be used within the tool?
e. Can different modes be integrated effectively separately into the design using timing settings?

A synthesis of these items suggests that the ability of the tool to facilitate particular modes associated with a given rhetorical effort and the timing of those modes in combination or separately within the space provided are affected by the tool's affordances and constraints.

Recalling Kress and van Leeuwan's (2001) point about design and production relative to certain media capabilities, Tufte (2003) also observed that some tools are better than others for communication related to engineering. For example, technical reports are better than PPT for conveying certain kinds of information. This is because content in a single PPT slide is affected by the size of slide and the fonts available. Also, bullet lists, a popular device in PPT, which is designed as a business/marketing tool, leaves causal information out. A popular example Tufte used to illustrate this was his analysis of the link between a presentation given by NASA scientists before the *Columbia* space shuttle disaster regarding a concern about tiles damaging the shuttle and the rhetoric associated with that presentation. He pointed out that the slide show associated with the presentation used PPT and that a reference to the concern was embedded as a secondary item in a listing; as such, it was in smaller text than other textual information on the slide. So certain media have the potential of limiting an audience's understanding of one's message.

Much as Moreno and Mayer (2000) identified attributes of multimodal rhetoric within learning contexts, Tufte (1990) helped identify attributes of visual representations that affect a reader's understanding of quantitative information. Combining effectively designed graphics with audio narration or print-linguistic materials contributes to the rhetoric inherent in multimodal representations of information. However, the medium's capabilities affect whether and how these can be integrated.

WRITING TECHNOLOGIES AND NEW MEDIA

People can create slide shows that integrate audio narration to describe processes or propose changes. Designers create videos, including those that use virtual environments, and place them on the Internet via YouTube for the public to find and view. However, these can also appear through links to websites. In this section, I describe attributes of each medium affecting design and cognition. The specific examples of each medium that I mention to illustrate the problem and related attributes are associated with subsequent chapters.

Cognition and PowerPoint Slide Show

Tufte's (2003) work with slide show designs and related rhetoric affects considerations related to slide shows, especially those involving PowerPoint. As noted, Tufte acknowledged that PowerPoint is designed as a sales tool, consequently, it makes extensive use of slide templates that include bullets and different font sizes relative to an item's position as a heading or subheading in a list. Tufte also noted that the more information presented on a single slide the more a viewer needs to absorb. A slide cannot be used as a regular page of text.

Another popular slide show tool is Prezi. One of the first major comparisons between PowerPoint and Prezi was that Prezi facilitated nonlinear presentations—one could move back and forth between slides easily. PowerPoint, though, could do so as well, so it was perceived to be more linear. However, this back and forth movement that Prezi facilitates can be unpleasant for the viewer if overused in a single slide show. Other contrasts between the capabilities of the two tools include that Prezi has a limited set of fonts while PowerPoint has considerably more to choose from, which can impact text design; and one can integrate more drawing tools with PowerPoint than with Prezi (Clark, 2013).

Tufte (2003) observed that the simplicity with which PowerPoint allows one to create slides encourages an outlining format that limits the amount of information one can present with each slide. This favors the rhetoric of a sales pitch—concisely presented and pointing out the most important attributes quickly. However, the same design hinders cognition within instructional settings. Teachers often use it within a lecture format to present information about a new concept, but it cannot facilitate cognition on its own without additional media or tools to provide additional information needed for learning.

Video Rhetoric

Another tool that is used in developing instructional materials is Movie Maker or similar video production tools. Some faculty develop a video lecture or use a video from the Web to help students learn. However, again, the tool's ability to integrate various modes and one's ability to use the tool affect design. Movie Maker and iMovie have their own strengths and weaknesses as video production tools. For example, iMovie allows for more tools to be used for viewing (export media) than Movie Maker, and iMovie is strictly for Mac machines while Movie Maker is for PCs, so the particular operating system one has greatly affects which tool one can use for video production. However, faculty use such tools depending on how they understand they can be used. Some have more experience with a given tool than others, and this affects how they design learning materials with them.

Johnson (2008) observed several questions that educational professionals face when developing a video product:

> What screen size should the videos be, what recording tool should you use, what microphone is best, how long should the videos be, what file size is acceptable? Should you use voice or captions? Where will you create the recording?
>
> You can create video tutorials using dozens of different methods. There are no official steps to create videos, because situations and audiences vary so widely. If you're creating e-learning with quizzes for a global audience, your approach will be different from one who is creating demo videos for a small company. (para. 1)

Composers also need to consider such rhetorical issues as camera angle, background, use of colors in text, and other images when they develop web texts that integrate video (Wysocki, 2001). While some of these are associated with reader-perspective assessment, the recording of any narration and camera angle are affected by the tools themselves in some settings. I tend to focus on transitions, camera angles, and integration of audio and images. However, I discuss attributes of each affected by the tool itself in another chapter.

Multimodal Instructional Theory: Narration/Images

According to Mayer (2001), multimodal presentation is most effective when learners with little or no previous experience with or knowledge about a task and/or those who have the ability to process spatial information quickly are involved (p. 161). When an audience has some experience with a given topic or concept, other approaches that involve fewer and different modes of representation can be used. Mayer acknowledged the importance of presenting both images and words in conjunction with each other. Such presentation helps the learner "to hold mental representations of both in working memory" (p. 96). According to Baddeley's (1986) model of working memory, there is a phonological (auditory) channel and a "visuo-spatial" (visual) channel associated with short-term memory. Schnotz (2005) suggested that when a visual image is presented to a reader, the reader can create a visual model as he/she listens to a narrative about the picture (pp. 54–55). By facilitating use of both channels, people can better process information than they can when too much of one system is used. However, the quality of the narration may be affected by the tool used to create it. In each case, though, design needs to consider the capabilities of the tool to facilitate certain content and cognition.

Multimodality and the Web

Web designing tools have developed to the point that most people are able to use even basic applications in Microsoft Office Suite, like Word or Publisher, to design web pages and websites. People can also use existing free templates to help create websites. In particular, page designs associated with technical communication texts and use of Web technologies and rhetoric associated with them are multimodal in many ways.

One example of a comparison between web-design and print-linguistic multimodality is the bomb manufacturing Standard Operating Procedure page I provided in an analysis of training materials at a munitions plant (Remley, 2009). I include that figure in a later chapter. The 2-page spread includes one side with only photographs showing what the step looked like, helping one visualize the step. On the other page is print-linguistic information set up under different categories so that a reader could locate particular information quickly. This is similar to web page design rhetoric of dividing a page into multiple columns or boxes and letting the user scan a page to allow for quick access of certain information he or she wants.

CONCLUSION

Given the impact that a tool's capabilities has on design relative to a given medium, assessment needs to explicitly integrate such considerations. Recognizing the differences in the capabilities of different media and different tools helps consider how they may be used to design instructional materials that affect cognition. Part of the rhetorical situation is choosing which medium or media to use to design a message. This decision frames the design of the message itself, consequently, this is an important framing principle of the model.

CHAPTER 5

Historical Case Study: TWI Training Practices

In the previous chapter, I described a number of ways empirical research can apply the new model, combining scholarship in multimodal rhetoric and neurobiology. In this chapter, I provide a more detailed analysis that applies the model toward a particular historical case study related to hands-on training practices and in which an accident promoted revisions to training materials. I start with a description of the accident that prompted changes in approaches and then discuss findings related to preaccident training and postaccident training to show the changes in the ways the modes were presented, modifying some aspect of the sensory experience and cognitive process. As I do this, I link the model to that discussion. This includes references to social science scholarship and neurobiological scholarship.

While this is a historical study, the training program is used in the lean manufacturing programs that are esteemed in industry today. Much as was the case in the previous chapter, I apply the model theoretically because it is a historical study and because no neurological data about the experience exists. However, much of what is understood about the effects certain practices and experiences have on cognitive development and cognition informs this application.

In particular, I consider the approach to training applied in the Training Within Industry (TWI) program at a World War II arsenal in the United States. TWI was developed in partnership between government and industry, and it was utilized at arsenals throughout the country. I provide a case study of how TWI was implemented in a particular arsenal, including some background related to the program. While particular materials and approaches were used initially, a catastrophic accident at the arsenal affected changes in the materials used in training. This case is presented within a different theoretical framework in another book (Remley, 2014). However, it is useful in showing an application of the principles of the cognitive model here. This case study provides information about the initial applications and the working out of some weaknesses in the multimodal rhetoric to facilitate better cognition.

As I pointed out in that previous publication, workers came from various backgrounds, and the TWI program attempted to address low literacy levels at the time and facilitate learning of how to manufacture war materiels quickly. However, there is evidence that a failure to balance effectively the literacy practices that were esteemed with those associated with the accommodations made for employees with low literacy levels contributed to a bomb accident that resulted in the death of 11 employees as well as destruction of a storage igloo and damage to several other buildings (Remley, 2014). Different neurobiological dynamics occur with different media to affect cognition, posing challenges in how best to present information for a given audience. Further, I describe changes made to print materials and other practices that suggest a refinement of the multimodal rhetoric provided in the materials and practices.

In this discussion, I apply an understanding of neurobiology as I apply the new model. This discussion illustrates the application of the neurocognitive model of multimodal rhetoric introduced in the previous chapter, showing connections and interrelationships across the various modes used in the TWI training approach and neurobiological dynamics associated with them. There are substantive differences in the way instructional information was provided before the accident and afterwards that demonstrate recognition of the attributes of the theory.

It is important to consider the various literacy practices that occurred at the Arsenal and implications thereof. Generally, a variety of literacy practices were used at the Arsenal, covering the range identified by the New London Group (1996): print-linguistic texts, aural, visual, spatial, gestural, and multimodal. Also, different forms of documents required different kinds of literacy skills, and I have noted previously (Remley, 2009) the appearance of a literacy-related hierarchy in which more print-linguistic skills are required of employees at higher levels. A study of the investigation report related to the accident offers insight into these practices as well as implications related to the separation of print-linguistic practices and oral/visual/experiential practices. A connection between literacy practices, training, and actual practice associated with this accident will be discussed throughout the chapter. I detail the methods that I used for the study in the other book, however, I present that information here as well, since the same methods can be used to collect information to facilitate study within the framework of this model.

METHODS

For this historical study, I relied on social science methods emphasizing content analyses. These included analysis of archived documents at the Arsenal as well as of interviews with several people who worked at the Arsenal from 1940 to 1960. The document analyses included readability testing, using the

Flesch-Kincaid grade-level test, and analysis of the modes used in documents. I discuss some challenges related to this study in Appendix A, which repeats some material from that previous book. I provide interview questions in Appendix B.

Much of the background information in this chapter is derived from the *Training Within Industry Report* (War Manpower Commission, 1945) and some historical documents from the period 1941–1945, as well as the *TWI Workbook* (Graupp & Wrona, 2006) and *Training Within Industry: The Foundation of Lean* (Dinero, 2005). The *TWI Report* document served as a primary source of information in understanding the specific programs that TWI implemented at plants throughout the country during WWII. Documents available at the TWI Service website were also reviewed, bridging some of the gaps between data points.

TWI BACKGROUND

The Training Within Industry (TWI) model was developed by the War Manpower Commission early in World War II (WWII) as a model to help standardize training and some operations around the United States during WWII. As I mentioned earlier, the philosophy developed by those who created the TWI program has been espoused in recent years as one that can lead to more efficient production methods, and it has been implemented in several companies (Liker & Meier, 2007; TWI Learning Partnership, 2015).

TWI includes three principle programs, all of which begin with the word "Job"; they are called the "J-programs." These are Job Instruction, Job Methods, and Job Relations. Job Instruction details how to train workers, Job Methods focuses on improving performance and processes through innovation, and Job Relations addresses management of conflict between personnel. A fourth program is associated with the newer iteration of the program—Job Safety. This information originally was integrated into the others, however, as job safety programs developed, it became its own program within the TWI model. Again, I focus discussion in this analysis on training.

Even before U.S. involvement in WWII, the nation was manufacturing materials to support the Allies in their efforts to turn back German and Japanese Imperialism. With his "Arsenal of Democracy" speech in 1940, President Roosevelt announced that he had authorized construction of several arsenals around the country that would develop munitions and act as warehouses for ammunition for the allies in the war. He framed this project around the fact that the world was facing a challenge to democracy from imperialist nations such as Germany and Japan. Generally, the speech encouraged industry to change toward a war economy and lifting many restrictions on production and hiring of workers (see Appendix C for a transcript of the speech).

In the early years of WWII, it was evident that the specialized production efficiencies used by industry in the 1930s would be replaced by the needs associated with a shortage of workers. Nelson (1995) indicated that, "World War II production relied heavily on additional labor and a more elaborate division of labor . . . most assignments could be broken down to the point that prior industry or product-specific experience was unimportant" (p. 142).

Labor Market

Many skilled tradesmen enlisted in the military, and many people who remained were more familiar with farming skills than with industrial skills. The Arsenal employed between 15,000 and 19,000 people during WWII, and many of these were migrants from West Virginia and the South. Prospective workers were bussed to the Arsenal for interviews and signing of employment papers and then moved on to training.

Government-Sponsored Training During WWII

Nelson (1995) stated that, as the government recognized the problem of the mismatch between workers' skills and job requirements, "officials launched ambitious remedial efforts that became highlights of the mobilization experience [and] . . . the retraining of workers proved to be comparatively easy" (p. 143). The War Manpower Commission, in conjunction with industry leaders, developed the TWI program, and representatives held conferences around the country to help munitions plants implement it. According to the Arsenal's historical documents, officials from the Arsenal attended a conference hosted by a local office of Training Within Industry.

TWI's founding principle was that "'What to do is not enough.' It is only when people are drilled in 'how to do it' that action results" (War Manpower Commission, 1945, p. xi). The most basic objective of the program was to train supervisors and workers quickly to produce defense materiels in order that such training would result in efficient and quality production. The report acknowledges that

> The training we give the worker . . . can be more than an expedient means of getting the job done. It can be suitable to the individual and in line with his native talent and aspiration. Then it becomes education because the worker . . . trained in accordance with his talent and aspiration, is a growing individual. (p. xii)

As with Mayer's (2001) observation about accounting for individual differences based on learners' experiences and knowledge prior to receiving instruction, one can apply the concept of prior experience to considerations of the motives for the TWI program. By emphasizing modes with which workers were already familiar in their learning and training, literacy problems could be minimized while facilitating cognition.

Also, Kress and van Leeuwen (2006) posited that production of meaning includes "experiential meaning potential." They stated that, "humans have the ability to match concepts with appropriate material signifiers on the basis of their physical experience of the relevant materials" (p. 75). The more experiences one has with a given set of procedures, the more he or she can understand certain routines. The "prior experience" attribute of the model is engaged here, particularly relevant to development of synapses based on how one learned in the past; training emphasizes an approach with which workers were already familiar through previous learning experiences—hands-on training.

TWI JOB INSTRUCTION

Job Instruction covers processes to help train lineworkers to perform a particular task and to train supervisors to perform their tasks. The TWI program broke down jobs into simple processes (War Manpower Commission, 1945, p. 21). After operations were broken down into simpler tasks, training on each task included the following form of instruction:

1. Show him how to do it
2. Explain key points
3. Let him watch you do it again
4. Let him do the simplest parts of the job
5. Help him do the whole job
6. Let him do the whole job—but watch him
7. Put him on his own (pp. 19–21)

Generally, the following four essential elements were presented to plants:

1. The training program should be one of utter simplicity.
2. It must be prepared for presentation by intensive and carefully "blue-printed" procedure, utilizing a minimum of time.
3. It must be built on the principle of demonstration and practice of "learning by doing," rather than on theory.
4. The program should provide for "multipliers" to spread the training by coaching selected men as trainers who, after being qualified in an institute . . . pass the program on to supervisors and their assistance who would use it in training men and women workers. (p. 32)

This training emphasized visual, aural, and gestural/behavioral (through practice). These attributes were also included in multiple levels of training. I described the program's original implementation previously (Remley, 2009, 2014). This included using films, narration, models, supervisory training "which included instruction on how to teach a job, management principles in regard to human relations, and later, instruction on how to improve the method of doing the job" (Remley, 2009, p. 104).

Again, presenting text and image simultaneously enables the learner "to hold mental representations of both in working memory" (Mayer, 2001, p. 96). According to Baddeley's (1986) model in *Working Memory*, there is a phonological (auditory) channel and a "visuo-spatial" (visual) channel associated with short-term memory. Schnotz (2005) suggested that when a visual image is presented to a reader, the reader can create a visual model as he/she listens to a narrative about the picture. So the program integrates intermodal redundancy well. By facilitating both channels, one can process information better than they can when too much of one system is used. Further, allowing the trainee to observe the process engages the visual dominance attribute of the model: One sees the entire task performed visually before any narration occurs. It also enables mirror neurons to gain some information concerning how to imitate the trainer.

Pinker (1997) explained that mental representations can occur in various modes: visual, auditory, print-linguistic, and "mentalese" (pp. 89–90). How one comes to their own mental map of a given concept depends on how that person has experienced a given phenomenon previously and is able to generate a mental abstraction of a concept. Mental maps help a person to understand new experiences through the engagement of prior experiences related to the new experience and relative to modes with which they have interacted previously relative to those experiences. Again, the program used a method of learning that the workers had used previously: application of prior experience to facilitate cognition is useful as well.

The documents associated with TWI's instruction and improvement models illustrate features of mapping using combinations of multiple representation systems.

Manuals

Once a worker completed training, they were very familiar with their job. Manuals were used primarily as reference materials. Generally, these manuals integrated text and graphics, which I detailed previously (Remley, 2014). While the standard operating procedure (SOP) for howitzer shells separated textual information from pictures, placing images in an appendix, the SOP for bomb production integrated the two more closely with each other.

One may assume that the entire operation may not have been effectively represented textually or through pictures or a combination. Amerine and Bilmes (1990) stated that instructions may assume a certain embodied knowledge possessed by the user. While information is not explicitly represented textually or graphically, certain attributes of the operation may be implied (p. 327). However, as Hutchins (1995) noted, one's relationship with the information affects what can be considered appropriate information to include or exclude. The workers in Hutchins' case study were technicians in a nuclear submarine, so they likely had more knowledge about nuclear equipment than one transitioning from farmwork would have about war-industry work. The prior knowledge of concepts associated with the nature of their work would enable them to negotiate information from their formal training with challenges they would face "in the wild." However, one moving from farmwork to war industry would not have a similar prior knowledge or experience with the particular work involved and would need more specific information as well as adhere more to specific instructions provided in training. It is this challenge to negotiate one's prior experience with learning new tasks that contributed to the accident in this case as well as prompting revision of training materials and approaches to call attention to particular safety protocol.

Prior to presenting the case study, I review the attributes of the model. Again, the attributes include

1. Intermodal sensory redundancy—the preference to integrate more than one sense to facilitate reinforcement of information and allowing mirror neurons the opportunity to gain information to facilitate imitation;
2. Visual dominance—recognition that the visual sense is usually engaged first, and as such is dominant in information processing;
3. Temporal synchronicity—the timing of exposure to different stimuli affects how related information is processed;
4. Prior experience—the role that previous experience with certain information or learning style has on acquiring new information; and
5. Attention-modal filtering—that one must filter certain modal information in an effort to concentrate and best process relevant multimodal information.

I briefly describe how each element of the new model can be applied to discussion of the TWI program.

The TWI program, as described above, implements multiple senses in a redundant manner. Between observing demonstrations, listening to a trainer provide instruction and performing the task, one experiences visual, auditory, haptic, spatial, and smell sensory experiences. Further, there is an emphasis on visual sensory information and modes. One sees a task demonstrated and film is also provided if necessary. While instruction includes narration, the Colavita visual dominance effect will occur given the presence of visual modes. Generally, the different senses are engaged at particular times in the training process, so it is possible to consider the role that temporal synchronicity plays in how one learns the task.

The TWI program was developed because of workers' low literacy levels and because of their existing background and training experiences. For the most part, this prior experience affected how those who developed the TWI program designed training to integrate similar modes and approaches. Finally, there is evidence of attention-modal filtering with respect to moving from the demonstration phase of training to performing the task themselves. As one observes the various steps through the first three parts of training listed above, one pays attention to particular information presented multimodally, and it is possible to study changes in that attention with each phase of the demonstration.

The accident that prompted revision of training materials illustrates implications associated with a lack of consideration to these principles, including the prior experiences of the audience, so I begin the case study with a description of that accident.

THE ACCIDENT

Description of Event

On March 24, 1943, an explosion in the depot area of the Arsenal killed 11 people and injured another 3. The explosion destroyed an entire storage igloo as workers were unloading and moving boxes of small, 20-pound fragmentation bombs (see page 85 in Remley, 2014). The investigation concluded that the combination of a defect in the fuse used to detonate the bombs and the rough handling by workers, in spite of expressed precautions that pertained to the defect, caused the explosion to occur (War Manpower Commission, 1945, p. 13).

Report of Investigation into Accident

In the investigation report, Stratton (1943) acknowledged various attributes of practices that depot workers used when unloading and moving bombs into position within storage igloos. This information suggests that employees received specific training but moved away from some attributes of that training in actual practice.

Based on interviews with workers, Stratton (1943) acknowledged the routine procedure for unloading the bombs and a change in procedure as more boxes were unloaded or as workers perceived the need to unload quickly.

> The routine procedure noted above follows. The railroad car is opened. The door bracing and bulkhead bracing are removed. The semi trailer is backed to the door-opening, and beginning with the outside box of the upper tier the boxes are handed down and placed in the truck. This first unloading operation requires the boxes to be handled by two men but as more boxes are removed from the car it was common practice for one man to pull a box from the tier: (the box and contents weigh 168 pounds), permit it to slide down through his arms on to the floor and then to walk the box on its corners to the truck, where it would be taken by two stackers and placed in position. This procedure was continued until a trailer load was completed. (p. 7)

Workers were trained to unload the boxes as a 2-man operation. However, the workers modified it so that only one worker would do the job of two, resulting in dangerous handling of the boxes. They evidently had discovered in the course of their work that they could manage the work more quickly if each person unloaded one box as described because of the fuse's tolerance. This raises a point about the prior experience principle; even after training, one learns certain procedure-related behaviors that may compromise safety under certain conditions.

Workers were told to be careful, but not specifically how to be careful. Stratton (1943) acknowledged that,

> The handling of the boxes at both the railroad cars and the igloo was carried out according to usual procedure. Helper No. 3070 working in the railroad car substantiated the previous testimony but indicated the boxes containing the bomb clusters were difficult to handle inasmuch as they were not provided with handles. He stated they were instructed to give special care to this shipment as it was the first shipment of this type of munition that had been received at the Depot. He detailed the procedure of unloading the cars as described in paragraph 5 above, making the statement that the boxes were permitted to slide down between the arms of the workers, landing on the end, and then were walked to the truck. (p. 11)

In spite of being "instructed to give special care to this shipment," the workers used a procedure other than the one associated with their training, which compromised safety. Management may not have known that they used such a 1-man protocol, since it was not part of the actual training. Consequently, assuming that workers used the 2-man procedure that they were trained to do, management would not have known to discourage workers from attempting the 1-man practice.

Much of the training at the Arsenal came in the form of intermodal sensory redundancy as workers experienced visual, aural, and experiential training, per the elements of the TWI description above. Workers were shown how to do a given task through demonstrations and then they practiced doing the task. In some cases, very little time passed between the start of training and the worker assuming work on the task. This may have affected attributes of temporal synchronicity within cognitive processing. Further, attention-modal filtering may have contributed to the accident. One does not know what distractions there may have been within the training. One also does not know what safety tips workers were provided with orally but that were not listed in documents.

Stratton (1943) also acknowledged that the fuse assembly of the bombs involved was new and defective in its tolerances:

> During the discussion, Captain Dorsey of the Kingsbury Ordnance Plant showed data which indicated fuze —110 had had an extremely bad loading history . . . over one million of these fuzes have had to be re-worked before

> they could be made sufficiently reliable to be placed on the bombs. The chief weakness in the fuze is in the pinion column which when coming out of adjustment generally permitted the safety blocks to fall out and the fuze to become armed. This data report in the form of a memorandum to O. E. Ralston dated January 6, 1943, indicating the defects in these fuzes, will be requested from Kingsbury Ordnance Plant. (p. 11)

This passage indicates not only that a defect that could endanger the lives of others was known prior to the accident, but that a memo—a print-linguistic document—was written by an officer who was aware of defects in the fuse of the particular bomb.

While workers received their training mostly through visual, aural, and experiential practices, administrators often communicated through print-linguistic forms of literacy. Workers also read print-linguistic materials, though literacy expectations across these materials and positions varied considerably. Training practices affect how employees perform certain tasks, and these practices include the various ways they are taught about performing those tasks. There were clear differences in literacy practices relative to one's position and relative to the accident's timing.

The information in the investigation report suggests that workers did not process information about safety protocol well, and it shows that workers were not told adequately how to be careful handling the boxes despite officers' knowledge about the need for such care. A memo that specifies problems with the fuse and related hazards existed before the accident, and testimony shows that workers were told to be careful. However, their use of the dangerous 1-man protocol suggests that they did not know exactly why they needed to be careful or how to be careful.

A breakdown in communication occurred somewhere between the original memo and the instruction to the workers to be careful. This breakdown suggests that certain information ought to be conveyed in multiple modes of representation—orally and in writing—to reinforce each other. Indeed, Stratton (1943) concluded the report with the acknowledgement that,

> Upon verbal report of the undersigned to his commanding officer, action was initiated to locate all other shipments of twenty pound fragmentation bombs —41 in transit or storage and to caution all authorities that extreme care must be taken in the handling of these bomb clusters in shipping boxes. (p. 14)

This cautioning of "all authorities," evidently included supervisors, because the Ammunition, General (1945), used by supervisors in their training, included explicit references to precautions associated with rough handling or dropping of boxes, which were highlighted with italics to call attention to them.

g. Fuzes.
(1) Extreme care must be taken in handling and assembling fuzes to shell or bombs. All fuzes must be treated as delicate mechanisms. the forces which arm a fuze on firing a weapon can be simulated by rolling or dropping, and a fuze so armed may be functioned by the impact of a blow or by dropping.

No such references occurred in manuals that predate the accident. That none existed prior to the accident and did exist subsequent to the accident suggests that developers recognized the value of including such information and reinforcing it visually with italics. Including such information and highlighting it with italics facilitated intermodal redundancy through print-linguistic means; highlighting the word calls attention to it.

MODES EMPHASIZED IN THE TRAINING PROGRAM

The ways that workers were trained and their related practices contributed to the accident. I describe rhetorical attributes associated with the modes of representation used, and I integrate what is understood about neurobiological processes to the analysis of this training and how training practices contributed to influencing the accident.

Training Practices: Preaccident

According to interview participants who worked at the Arsenal and Arsenal documents, virtually all employees experienced some part of the training program, no matter the level at which they worked. However, training of lineworkers differed from that of supervisors, and training materials illustrated some differences in assumptions about cognitive abilities across those populations—lineworkers versus supervisors.

Lineworkers

Training of lineworkers emphasized hands-on approaches while minimizing the need to read manuals.

Roger (a pseudonym), the interviewee who began work at the Arsenal the earliest (1942) reported that training was heavily visual and hands-on:

> Me: Do you remember the training that you received when you begin, first began work at the arsenal
>
> Roger: Very little.
>
> Me: Very little. Do you remember? Very little recollection of it or very little training?

> Roger: You pick up a shell, put it in a vise.
> Me: And they showed this to you, or you didn't have anything to read but they showed you how to do this?
>
> Roger: No, no. It was very . . .
>
> Me: Hands-on?
>
> Roger: Hands-on. Very hands-on.

Training of lineworkers emphasized visual, oral, and experiential modes. I asked Roger specifically if there was any reading of print materials in this training, and he acknowledged that there was not. This is confirmed with review of archived documents, which also shed light on the training of supervisors.

According to the Arsenal's 1942–1943 Summary of Operations, the training programs offered used multiple modes of presentation. Principally, the training programs relied on visual modes, although manuals were distributed to employees. Mayer (2001) observed that multimodal presentation is most effective when low-knowledge learners (those with little previous experience with or knowledge about the task) and/or high-spatial learners (those who have the ability to process spatial information quickly, also visual learners) are involved (p. 161). The Arsenal's History of Operations (Atlas Powder Company, 1944a) acknowledged that,

> Training was concentrated upon two principal centers of activity:
> 1) employee training, which included an induction talk to all employes and pre-employment or vestibule training, lasting two or three days, for fuze and detonator line operators.

The induction talk was given by the employment interviewer. "Induction talks were given to over 19,000 employes from December, 1941 until April 1942" (Atlas Powder Company, 1944a, p. 110). As seen in the excerpted description below (taken from page 110) vestibule training included the showing of silent movies that were accompanied by narration (see page 89 in Remley, 2014).

Mayer (2001) acknowledged that it is important to present both pictures and words simultaneously rather than in succession. Presenting them simultaneously enables the learner "to hold mental representations of both in working memory" (p. 96).

Roger's comment and this passage indicate that training of lineworkers emphasized visual, aural, and experiential modes. Training for those who worked in the depot area also emphasized visual, aural, and experiential modes as well. The same document acknowledged that,

> Demonstrations as to the proper method of storing ammunition in magazines and of loading and blocking it in railroad cars were given by the instructors. In order to carry out these demonstrations miniature models of igloos, railroad cars, and several types of ammunition were designed and constructed to scale, thus making it possible to follow the specifications as given on the loading and storage charts. After the demonstrations had been completed, students were given ample opportunity to inspect and, later, to practice with the miniatures. (1943, p. 270)

Visual, aural, and experiential modes of presentation are evident in the description of the instruction. It relies heavily on the visual, though, engaging the visual dominance attribute of the model. Moreover, the worker was learning how to do a specific task in the context that he or she would be doing it. This kind of intermodal sensory integration combines the various modes the worker would experience in really performing the task, including smell and spatial orientation.

As reported by Bremner and Spence (2008), Lewkowicz and Kraebel (2004), and Keetels and Vroomen (2012), instruction that integrates multiple modes and multiple sensory experiences tends to facilitate better cognitive development relative to particular tasks. Further, the emphasis on visual modes reflects Wallace et al.'s (2004) finding that visual stimuli are important to facilitate learning. The more stimuli there are integrated, including visual stimuli, the more efficiently one can learn. Further, it echoes findings that integrating haptic sensory experiences (touch) relative to hands-on learning can enhance learning performance of tasks (Lacey & Sathian, 2012; Sathian et al., 2004).

Further, because steps are broken down and tasks are learned through repetition of demonstration, narration, and performance, this method facilitates attention-modal filtering. One does not need to be aware of more than what is going on with a simplified task. Also, because narration occurs after the demonstration, there is an ideal temporal synchronicity. Audio information is provided subsequent to and then with demonstration.

Finally, the training was in a similar way that workers were used to being trained—hands-on techniques. This reinforces what is understood generally about the influence of prior experiences in learning (Gee, 2003; King et al., 2004; Lacey & Sathian, 2012; Moreno & Mayer, 2000; Pinker, 1997; Wallace, 2004).

While lineworkers' training emphasized hands-on approaches with very little opportunities for reading manuals, supervisors' training included using manuals as well as classroom instruction. The next section describes more of the training program as it was provided to supervisors.

Preaccident Training—Supervisors

The TWI program identified five particular needs of supervisors: knowledge of work, knowledge of responsibilities, skill in instructing, skill in improving methods, and skill in leading (War Manpower Commission, 1945, p. 48).

While lineworker training lasted only a few days, supervisor training lasted 8 weeks, considering the time needed to train in the five needs areas identified above. Also, while most of the instruction portion included explanation of visual diagrams and demonstrations of production procedures, half of it (144 hours) involved fieldwork. Relative to the time spent in fieldwork, the Atlas Powder Company (1943) explained that, "it was felt that this was a minimum in which they could absorb, through actual handling, sufficient knowledge of the methods and procedures involved in handling the enlarged variety of ammunition now is use" (p. 267). Also, storage charts were made available to and studied by trainees prior to their going on to fieldwork (p. 273). Such charts emphasized visual literacies.

Once again, there is intermodal sensory redundancy in terms of trainees experiencing information through lectures (aural) and reading print materials that included some visuals (visual) and then moving on to fieldwork including demonstrations and practice with models—visual, aural, haptic, spatial, and smell. As mentioned in Chapter 2, combining visual and auditory information facilitates faster processing of information generally (Gutfreund & Knudsen, 2004; Kajikawa et al., 2012; Kayser et al, 2012; Mayer, 2001; Moreno & Mayer, 2000; Munhall & Vatikiotis-Bateson, 2004; Newell, 2004).

Postaccident Training—Manuals/SOPs

While there does not appear to have been a change in the visual, aural, and experiential approaches used in training after the accident, there was a dramatic difference in the print-linguistic materials used with that training after the accident. Generally, print materials published after the accident integrated far more visuals than print materials published before the accident, engaging visual sensory experiences and facilitating more mental mapping of the processes. As I described in "Exploding Technical Communication" (2014), the 1942 manual included very few visuals while attempting to describe processes textually. Conversely, the 1945 manual integrated images of two different kinds of ammunition, helping the reader visualize the particular item being described by text. Pages 98-99 in "Exploding Technical Communication" how a 2-page spread taken from a bomb manufacturing SOP published in 1945. Mayer (2005) encouraged providing images and textual information together so that each reinforces the other. Mayer's multimodal principle is that people learn better when pictures and words are integrated into an instructional message than when only words are used (p. 63). Again, this is an application of intermodal redundancy as well as the visual dominance attribute.

When only words are used, people may attempt to "build a visual model," but they may not attempt to do so. If a picture is provided, people can make the visual connection more readily. Note also the print-linguistic text regarding safety positioned to the left, among the first items one would read when reading a traditional print-linguistic text from left to right. The headings related to each section also suggest a means to accommodate modal-attention filtering; all of the relevant information related to a specific attribute of the step can be focused on with the use of the sections. Finally, even the print-linguistic text in the supervisors' manual called attention to safety, including the handling of fuses and issues with dropping packages. Not only was the information provided explicitly, but it was italicized. The visual feature of using italics to highlight the text called the reader's attention to it. This is an example of the visual dominance and intermodal redundancy principles of the model. The use of graphic visual representations in addition to the print-linguistic text encourages the reader to look at the image and absorb information provided there, reinforcing the print-linguistic text.

These examples suggest that in addition to using print-linguistic literacies in training after the accident, more effort was made to integrate visual attributes into print materials to appeal to the visual literacies of workers and to reinforce the visual aspects of the experiential and aural training. The highlighting of certain text related to safety and positioning of it in certain places on a page contribute visual attributes to the print-linguistic text too, reinforcing the message. Once again, a relationship between visual processing and the Colavita visual dominance effect appears. Visual information facilitates mapping and understanding what a given process looks like relative to the objects involved. Using only print text requires one to imagine the operation. Emphasizing print text also goes against the prior experience attribute in the model and generally recognized in scholarship in both social sciences and neurobiology.

Changes to the manuals after the accident facilitated better processing of the information by including useful visuals showing specific spatial attributes of the objects involved and highlighting information visually. Hutchins (1995) indicated that print-linguistic instructions tend to act as a guide more than as a standard for all settings in which a given task may occur. This may be related to the habit of workers to use the 1-man protocol in handling bombs.

While their training included practice with the 2-man protocol, workers began using the 1-man protocol because they considered, at some point, the training they received to be guidelines. As they wanted to speed up the pace of the work, they came to believe that the bombs they handled could tolerate being dropped and handled roughly. However, the defective fuse did not have the same tolerance, and workers were not aware of that. Consequently, they continued to use the rough handling associated with the 1-man protocol, contributing to the accident.

Hutchins (1995) reported about operations in a nuclear submarine, documenting field practices that did not follow training or documented instructions and how such practice became accepted behavior. He called this practice "situated cognition." Hutchins explained that a set of documented procedures acts as a "meditational device" between the task and the reader, however, he also observed that learning can "be mediated by so many different kinds of structures" (p. 291).

These structures include visual/experiential modes. Hutchins characterized written instructions as guidelines, and Amerine and Bilmes (1990) pointed out that written instructions serve as a set of guidelines, and authors of instructions often omit information that they feel the reader may be able to infer. Hutchins further observed that reading instructions and performing the steps represent different mediating structures for the learner; the act of reading a single written step in a procedure involves understanding "what the step says, what the step means, and the actions in the task world that carry out the step" (p. 301). Hutchins explained that written instructions act as a guideline around which actual performance in doing the task may be negotiated because of the potential to infer actions from written instructions that may include limited information.

Again, Arsenal workers were not familiar, generally, with munitions prior to their work at the Arsenal, however, Hutchins' (1995) subjects were trained nuclear submarine specialists. As specialists, they would have specialized knowledge of particular operations and would not need highly prescribed instructions in training materials. Arsenal workers needed more explicit information, as they were provided in texts published after the explosion.

CHANGES IN TRAINING

Again, no accidents of the like occurred after this event. As indicated earlier, the author of the report emphasized acknowledgment of the need for caution with the fuses, and that acknowledgment was evident in subsequent manuals. Additionally, a few interview participants who worked at the Arsenal also recalled an emphasis on safety. Indeed, no other accidents involving such explosions occurred subsequent to this accident, suggesting that these safety precautions were clearly articulated in writing and reinforced several ways. Also, a number of interview participants who had worked at the Arsenal acknowledged an emphasis on safety.

David, who worked there after the explosion, acknowledged an emphasis on safety protocol among lineworkers:

> Me: What kind of training was there associated with that line work that you remember?
>
> David: I had some . . . there wasn't any training outside of safety. They had like a change house and on the actual line as far as pouring the melt. I didn't do this, but, on the line where they melt the TNT; that is not the name of it, but that's what it is; okay . . . as for safety we were not allowed to have any matches or, if you . . . like . . . there had been a change house a short distance away probably about . . . probably about 300 feet, 400 feet something like that; and that's where they allowed the people to smoke. No lighters and no matches were allowed in there at all. But they had a small lighter that was

> connected to electrical. And if you tilted it a little bit, it would light for the people who wished to smoke there.

Don, who also worked there after the accident, also articulated the existence of a manual that emphasized safety:

> Don: I was almost guaranteed a job before I ever got out there. It was in safety and security, which was a field I knew nothing about. But nonetheless, there was a field manual that you carried with you and that manual had been written in blood; and so you didn't vary too much from what was in the field manual.

David pointed out that training emphasized safety information, and Don acknowledged the existence of a safety manual that was treated as very important. While Don's position was directly related to safety, the acknowledgment of the manual and its importance suggests that workers were encouraged to use the manuals. Combining the visual (print-linguistic) with audio (oral instruction about safety) modes of representation in a way that reinforces each other is a form of intermodal sensory redundancy.

CONCLUSION

This case shows how the TWI program integrated all elements of the neurocognitive model of multimodal rhetoric and illustrated application of the model in a historical study. The application emphasized hands-on learning approaches, but it also included discussion of print-linguistic learning materials and how multiple modes were used therein. Further, it showed how rhetorical attributes of the instructional materials affected cognition before the accident and how the changes in materials after the accident improved cognition by more actively engaging the attributes of the model.

CHAPTER 6

New Media Applications: Slide Shows and Simulators

In the previous chapter, I detailed an application of the integrated model of cognition and rhetoric related to hands-on training and instruction related to a specific program at a given workplace. With the proliferation of new media and accessibility of it via the Internet and affordable technologies, many are developing instructional materials with digital tools that engage multiple senses. Slide show tools are used as visual aides within lectures and other learning environments, and simulators enable learners to interface with the particular environment in which they will work. Many of these applications involve training and learning toward helping someone understand information new to them. In this chapter, I provide further analyses applying the model relative to these two particular kinds of new media applications popular in education and training—slide shows and simulators.

I have used the particular cases I present in this chapter in previous publications that each had a different focus—production and assessment of them (Remley, 2010) and their use within a form of literacy sponsorship (Remley, 2014). New media integrates a variety of modes of representation in digital forms, providing multisensory experiences such as combinations of audio, video, and representations of spatial relations, so I apply the discussion of the new model to them here.

For example, scholarship in multimodal rhetoric has considered the effectiveness of new media such as instructional video, PowerPoint slide show software, and virtual environments relative to their ability to facilitate learning and training. I describe specific applications of the model to simulators in this chapter to illustrate how the principles apply to these media.

All of the technologies involved in simulators discussed in this chapter integrate varying degrees of virtual reality environments. Such environments attempt to engage as many modes of representation and related sensory experiences as possible toward helping one understand how to perform a given task under certain conditions. As such, there is considerable scholarship in both the fields of multi-modal rhetoric and biological neuroscience that examines the use of simulators for cognitive tasks and cognitive development. The sensory experiences associated with each affects how close to an actual hands-on experience one has at that point in the learning process. While instructional materials and learning experiences may go beyond the simulator technology, I consider the particular multimodal rhetorical and neurobiological dynamics at work within the simulator experience.

Studies are considering various attributes that affect the ability of a video product to facilitate learning and training. Video is used extensively in such cognitive tasks and integrates visual as well as audio and spatial modes of representation. One is able to see a demonstration of a given task, and that helps one understand to some degree the task and how to perform it. However, various attributes of design affect how well video facilitates cognition relative to a given purpose. I describe application of the model to a video derived from one of the simulation tools I discuss as well.

In each case, I am limited to theoretical discussion of these dynamics, not having used any of the data-collection methods associated with tracking neural activity such as EEG or fMRI or 2-photon microscopes. The analysis tends to emphasize the social science/rhetoric attributes as well. However, I also indicate what kind of analyses may be appropriate with the information I have.

Slide shows that include audio narration facilitate visual and auditory modes of representation and stimuli experiences. I provide discussion of a particular application of the model to slide shows to show the emphasis on these particular modes and stimuli. However, simulator technologies can integrate more modes and stimuli. It is possible, for example, to mix media to integrate video into a slide show.

SLIDE SHOWS

Slide show technology provides a way to present information for an oral presentation, which typically includes audio as well as visual modes of representation. However, it also enables one to create a multimodal piece that includes audio narration for play at a later time. In this section, I describe some attributes of cognition related to slide show technology. As mentioned in previous chapters, Tufte (2003) argued that PowerPoint limits cognition because of the way it is designed as a sales presentation tool. Scientific presentations are compromised because PPT's design encourages the user to prioritize information such that supporting information about a given point may appear unimportant. Scholarship in assessing slide shows facilitates discussion about cognitive aspects of slide shows. There are not many studies focusing on neural processes pertaining to cognition using slide shows, but I attempt to connect the sociobiological model of cognition to analysis of a few slides that revolve around a single topic.

As I have noted previously (Remley, 2010, 2012), many rhetorical attributes of traditional print-linguistic texts also apply to multimodal products. Borton and Huot (2007) observed that "all multimodal composing tasks should be aimed at producing effective texts appropriate for a specific purpose and audience" (p. 99). Odell and Katz (2009) observed that authors must consider rhetorical contexts with either kind of composition (p. W201).

Stern and Solomon (2006) acknowledged the importance of focusing assessment on a few important areas of the product—"those tied to the student learning goals for the paper assignment" (p. 26).

In an effort to limit the number of considerations for analysis related to cognition and application of the model, I focus attention on the following: application of narrative with slides, use of space on slides, content, and transitions from one slide to another slide. Such assessment criteria use a few traditional attributes of composition that can affect cognition while also including an important element in multimodal instructional rhetoric: combining narrative and visuals that directly impact cognition (Moreno & Mayer, 2000).

Tufte (2003, 2006) and Walbert (n.d.) identified a few rhetorical attributes of PowerPoint as a presentation aide, both indicating its simplicity as a weakness. Tufte (2003) pointed out that it was designed principally as a business presentation tool, its slide templates facilitating quick information using bulleted points and highlighting certain attributes of a message with larger font sizes. Design decisions associated with this templated simplicity impact cognition.

Walbert (n.d.) used the ability of a slide show to facilitate an understanding of information as a criterion for assessing slide shows (p. 13). His rubric for assessing multimedia projects included consideration of how the design of the slide show enables one to understand information. This consideration of design includes attributes of intermodal sensory redundancy and visual dominance, especially. Depending on the media used to include narration, it may also include the attribute of attention-modal filtering.

The following slides pertain to information about the kinds of composing nurses do in the course of their work. It was in response to a particular assignment in a sophomore-level writing course. The assignment asked students to research forms of composing that professionals in their major field do. Three students who were nursing majors responded, and they described various forms of composing nurses do, including the use of computer-mediated communication (CMC). One focused on texting and instant messaging, another focused on e-mailing, and the third described a range of genres used in various fields, including explicit references to nursing. I described attributes of particular slides within each slide show, with student permission, and criteria associated with their assessment and connections to multimodal theory and the new model.

Walbert (n.d.) argued that PPT is a visual aide and as such should supplement or complement the narration; the slide show cannot act independently from the narration, and any narration that occurs with the slide show ought not simply repeat text on the slides as if the speaker is reading a slide. Walbert referred to the rubric developed by the University of Wisconsin-Stout (2010) for guidance with design issues. These included textual items such as size of text and amount of text on a given slide (too much is difficult to absorb) and layout, which pertains to use of space on the slide generally. Such attributes can affect cognitive processing of information relative to filtering of information.

Images provided on slides should be an appropriate size, and audio should help explain the images and help the information to flow smoothly. Mayer (2001) explained too that the integration of narration and image consecutively and in conjunction with each other help to reinforce each other. Baddeley (1986) identified a phonological (auditory) channel and a visuo-spatial (visual) channel associated with short-term memory, both of which facilitate "working memory." Schnotz (2005) explained that the reader can create a visual model as he/she listens to a narrative about the picture when a visual image is provided as well (pp. 54–55). By facilitating use of both channels, people can better process information than they can when too much of one system is used. This considers the principle of modal filtering by reducing the amount of filtering needed.

Emma (a pseudonym) developed a presentation concerning a range of writing that various professionals do, and she referred to a couple of documents specific to nursing. The slide below illustrates a particular form that nurses must complete on a very regular basis (Figure 6.1).

The entire slide shows the complete form, and each portion of the form is noted in green text. This helps the viewer focus on the specific parts of the form as she describes it. This is a very good use of space, as she shows the entire form on a single slide and still has room for the annotations. The display of a single form helps the viewer to visualize the entire form, while the color coding of the section labels helps with attention-modal filtering. One can follow the narration easily referring to the labels.

Showing the specific form contributes to cognition because it provides the viewer with specific information she is likely to see in a workplace environment before actually having to engage with it. As such, this slide considers the prior experience attribute of the model. While the viewer may not have prior experience with the form when she views the slide show, this information can be considered prior knowledge when a viewer sees this information again in a subsequent nursing course.

Emma also included a slide on which she listed the various sections of a form called "Nurses Notes" (Figure 6.2).

Providing a listing helps to capture the essence of the content of the form, helping one process the information quickly. Much as flash cards facilitate quick learning through providing concise information about certain concepts, a listing of key items on a slide calls attention to those items and enables the viewer to recall the items. The narration associated with this slide also reinforced the

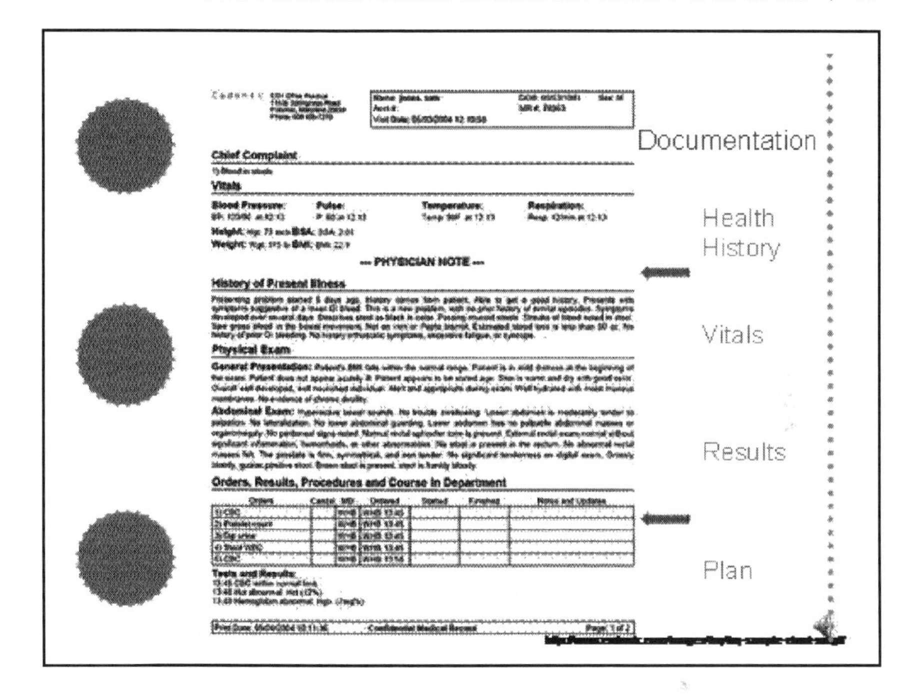

Figure 6.1. Example of form in nursing field.
All slides used with permission of students.

listing by describing more of each item. So these suggest the principle of modal redundancy as well as visual dominance; the information is visible as the viewer listens to the audio narration, and it reinforces the audio information.

Slide show technologies provide a limited space for information, but they usually involve the inclusion of audio narration, either live or digitized. These modes can act to reinforce each other, integrating the principle of intermodal redundancy. However, the slide space must limit the amount of information provided so as to avoid problems related to filtering modes. Narration that reinforces information in a pie chart can facilitate cognition better than if the pie chart were explained with print-linguistic text on the same slide.

SIMULATORS

While slide shows facilitate a lecture-oriented form of instruction, simulators allow users to interact with a replication of the actual environment and context in which a particular activity would occur, similar to the hands-on learning experience described in Chapter 4. Consequently, there are a number of multimodal attributes at work to facilitate cognition. Because the user sees things in three

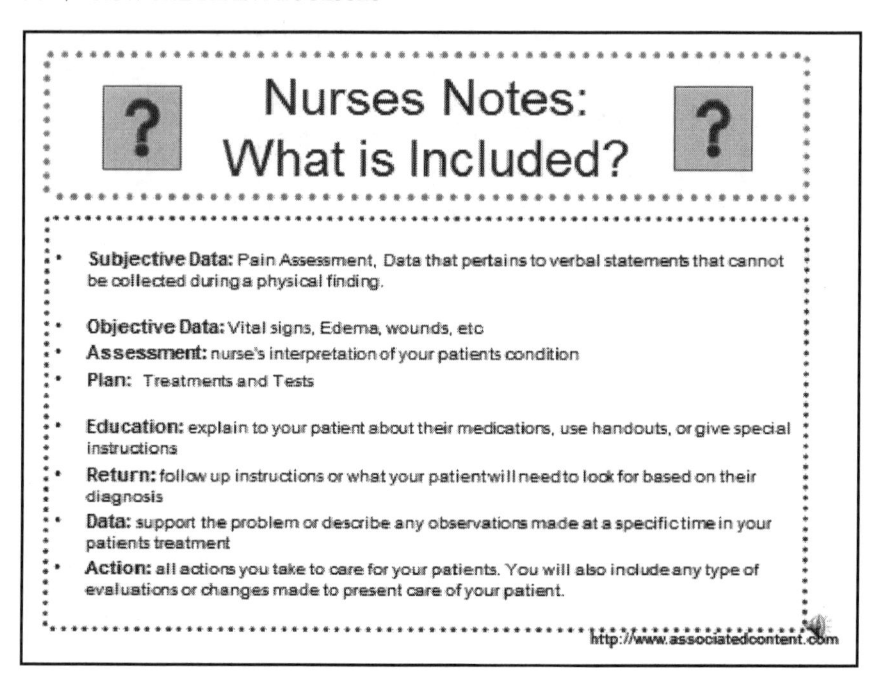

Figure 6.2. Nurses notes.
Used with permission of student.

dimensions, approaching the way one would normally see them, it facilitates visual and spatial modes. Because the user engages specific objects and tools, he engages tacit, behavioral attributes as well. Generally, an instructor is nearby to provide information orally, or audio narration is embedded in the particular digital application, so there is an aural attribute as well. If the simulator pertains to a particular screen that includes print-linguistic labels, one also experiences print-linguistic text.

Analysis of cognitive activity occurring within a simulator depends heavily on the nature of the simulator, especially with respect to its field of vision and its ability to synchronize visual information with motions. As Campos and Bulthoff (2012) indicate, most virtual environment applications involve a "stationary computer monitor paired with an external control device that is used to interact with the VE (i.e., a joystick or a mouse)" (p. 605). Such equipment has a limited field of vision, which consequently affects immersion. As the field of vision increases, one may perceive that he or she is more fully immersed in the scene or situation. If one is able to see a wall behind the display, however, that can limit the perception of the reality and that can negatively affect other dynamics associated with the experience.

Also, as Reid (2007) observed, when one's movements are not well synchronized with what they see and perceive, they can experience motion sickness; there is a "disconnection between various sensory inputs into the body: for example, our eyes register movement but our inner ears do not" (p. 94). He acknowledged that Heim (1998) referred to this phenomenon as "Alternate World Syndrome" (AWS), which involves the experience one has when switching between virtual and real worlds (Reid, 2007, p. 94). To explain it further, I relate my experience with a race car simulator at an arcade. The "game" involved sitting in a seat while facing a display that appeared to simulate a race track. The user's vision was positioned from the seat of his own car, and it included the ability to see in front of him and a limited amount to either side. Anything behind that peripheral field was a blind spot. As the vehicle moved in the display on the machine, one experienced this field of vision at a particular rate of speed, which was controlled by the gas pedal. Further, the seat moved so as to attempt to simulate what the driver may experience in the vehicle itself. While one drove, they would experience motion senses associated with balance and move in their seat accordingly. However, within about a minute of starting to play this game, I experienced motion sickness. The visual display and the seat motion were not well synchronized, so my brain experienced that disconnection of sensory stimuli associated with AWS.

Heim (1998), however, noted that we may be able to overcome AWS with experience. As our brain learns how to process the disconnection between sensory inputs, we may be able to avoid the nauseating experience of motion sickness in such interactions. This would involve plasticity dynamics to accomplish that balance. As the brain learns how to account for such sensory disconnections, new synapses are developed, changing the composition of the neural system.

In the next sections, I describe applications of the integrated model to particular simulator interfaces to facilitate an understanding of how it may be applied. Again, this description is limited because of the absence of neural data. Also, it lacks the specific human feedback one needs to perform such analysis. However, the descriptions attempt to help the reader understand how one may use the theory in a more empirical, data-driven analysis.

Second Life

A popular simulator for education in general and, increasingly, professional education, is *Second Life* (SL), a 3-dimensional virtual environment in which users can interact with objects similar to a real-life experience relative to space and visual information. A user creates a representation of him or herself (an avatar) and navigates the environment and interacts with others through that avatar. One is able to make their avatar walk or run, make various gestures and facial expressions, and there is audio capability, so one can "talk" through their avatar as well.

Mollman (2007) described an SL application used by Orkin to train workers to inspect homes for pests. An avatar enters a home and moves about it checking in particular places for evidence of pests. Hudson (2010) described an application related to training students to perform border inspections within a justice studies program at a college near the U.S.-Canada border. He stated that students using the SL simulation did 39% better on specific performance measures than students who did not use the SL simulation, suggesting that the simulation contributed to better learning of specific skills (p. 117).

SL has become a popular space for medical training. Mesko (2008), for example, described a space in SL wherein medical students could engage in a realistic patient visit experience. Medical students entered a facility and asked to see a patient from a list of available patients. After the student's avatar washed his or her hands at the in-world wash station, the student then was shown the patient's medical chart and had to respond to a few questions about the chart and make preliminary diagnoses at a nearby file cabinet based on the information on the chart. Figure 6.3 shows what the patient's room, including the wash station and file cabinet, looks like. The patient's bed is opposite the file cabinet.

If the student passed that test, the student was able to request particular tests to further understand the patient's condition.

Figure 6.3. Imperial College *SL* facility.
Used with permission by Imperial College London.
Island objects designed by Maria Toro-Troconis, E-Learning Strategy and Development Manager, Imperial College London.

One can analyze cognition associated with such simulation relative to a number of attributes within the model. The most evident one is the attribute of intermodal, sensory redundancy. The student interacts with various modes of representation that attempt to reproduce a scenario he or she would experience in reality—patient and materials he or she would use in practice. A study could ascertain how well a student learns how to interact with the space and information once he or she has experienced the interface a few times. Consequently, it may also facilitate analysis of prior experiences and plasticity. One may study, for example, how neuron composition changes with each experience with the simulation and track such changes toward ascertaining an optimal amount of training needed in the simulation before moving the student on to an actual clinical setting.

Object-Specific Simulators: Pilot and Air Traffic Controller Training

Simulators include police training simulators, mining simulators, and flight simulators (Figures 6.4 and 6.5), all of which allow one to interact within an environment that simulates one in which they would be working as they train. Such a simulator facilitates learning by encouraging the user to experience the kinds of senses associated with the task, such as touch and movement or behavior, that cannot be simulated in tools like SL. Students training to become pilots or air traffic controllers use such simulators often in their education, and I provide a description of certain attributes associated with this use and how the model may be applied to multimodal-neural cognitive analysis with such tools.

Pilot Training

Figure 6.4 is a simulator that helps students learn how to control a plane in various conditions, while Figure 6.5 pertains to another simulator that helps students learn landing approaches. With the approach simulator, students also receive a graphic printout of the path they took, which is represented on a flight map. With such simulators, one uses specific tools that one would use in real life, and the simulators are able to show how certain sequences would affect an outcome, desired or undesired, so the user can see what would happen without a catastrophe occurring. Students receive about 30 hours of training with the simulators, in addition to receiving instruction in the air with a pilot instructor, before being placed in a control position in a real airplane.

An annoying feature of the simulator in Figure 6.5 is that the view of the runway that is represented is actually a projection on a screen in front of the cockpit object. Consequently, while it reproduces an airplane cockpit and view from the cockpit, one can easily see a wall of the room in which it is situated to the left of the projection screen (arrow). So analysis of attention-modal filtering is ideal here.

Figure 6.4. Simulator view airplane.
Dirk Remley.

Figure 6.5. Interface of second airplane cockpit simulator.
Dirk Remley.

Likewise, the window providing a field of vision in Figure 6.5 is very limiting. This does not represent what one would actually see in a cockpit, but it focuses attention on instruments and learning how to read them. So, while the simulator limits the sensory experience, it helps to focus attention on a particular element of flying—understanding how to read and use the instruments in the landing phase.

Research about cognition with the model here would include analyzing how well the tool facilitates focus on the task and avoids creating a distraction due to the limited external (to the cockpit) visual information available. As such, one could analyze it relative to intermodal sensory redundancy, previous experience, and attention-modal filtering.

Air Traffic Controller Training

Air traffic controllers (ATCs) need to learn how to monitor aircraft as they fly, entering and leaving particular air space. Most of that work is done via visual mapping on computers facilitated through global positioning system (GPS) tracking devices. Air traffic controllers also help pilots understand when runways are clear for takeoff or landing. As such, they have to learn how to monitor traffic patterns on the ground around an airport—runways and roads around them.

Another example of simulators is new simulation technology for training air traffic controllers. Many ATC programs break down the responsibilities associated with air traffic control into different lessons to facilitate quick training. Further, simulators of varying types engage students in hands-on learning. Analysis of training with such a simulator can include attributes such as intermodal sensory redundancy, previous knowledge, and attention-modal filtering. In particular, how does one filter information from outside the screen to focus on the image on the screen? Also, how does one's previous knowledge of reading tracking tools help one learn and understand the tracking of aircraft?

Figures 6.6 and 6.7 show the actual landing and takeoff operations from the control tower's tools.

This is an overhead view of an airport and its runways. Lights representing aircraft would appear on the screen; there are not planes on runways in this particular image. However, a visual of the airport's runways is coupled with this screen, as shown in Figure 6.7.

With the experience, students get two different visual representations of the runways and can monitor aircraft moving about the runways accordingly. As such, analysis related to intermodal sensory redundancy is ideal. One can also consider attention-modal filtering dynamics to understand how students move from one representation to another to best understand what is happening, that is, which representation seems valued at different times in the monitoring processes and how one learns how to monitor aircraft on the ground. Also, each provides a different visual experience, so one could analyze processes regarding visual dominance elements relative to a given task or learning how to use a given visual tool.

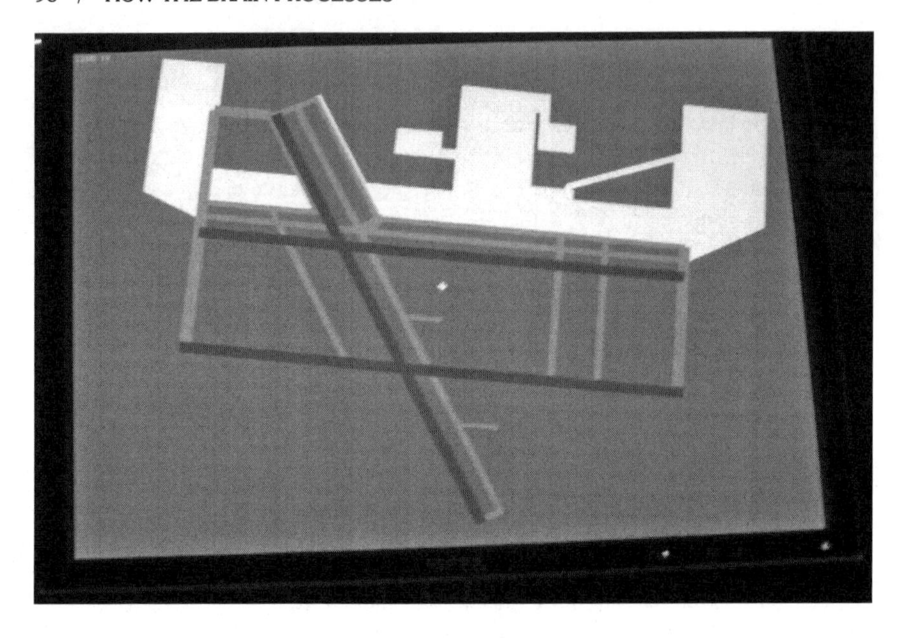

Figure 6.6. Overhead of runways.

Figure 6.7. Combination of computer display and tower simulation view.

In each simulation case, though, the principle of learning-by-doing applies, and this principle engages experiential learning as well as visual, aural, spatial, and behavioral modes of representation. Neural processing of such training is enhanced by what Gallese, Eagle, and Migone (2007) called "mirror neurons." In studies with monkeys, they found that when a monkey observed another monkey performing a particular action, the same neurons associated with performing the act were stimulated in the monkey observing the action (p. 132). Further, they cited several studies in which mirror neurons are observed in humans (p. 134). While they noted that the neurons became stimulated, suggesting cognition of the action, they also noted that such firing of neurons does not suggest that one understands the context in which the action occurs (p. 136). However, this information suggests that one may learn to perform certain tasks in context if repeatedly shown the motions/performance within a given context.

Simulators are used in several kinds of contextualized training, and multiple sets of neurons may be involved in processing information in such training toward cognition. Attributes of these are at play with other forms of new media as well, such as video products showing one how to perform a task.

CONCLUSION

Simulators and slide shows are used in a variety of training and education programs. The model provides a means by which to analyze the cognitive attributes of such tools and can help to improve instructional materials related to slide shows and simulations. The principles of the model apply not only to hands-on training, described in the previous chapter, but also to training that occurs with new media. This chapter focused on two forms of new media. New media also includes video, and the next chapter details some applications of instructional video.

CHAPTER 7

Instructional Machinima

In the previous chapter, I described some applications of the model to 3-D and virtual simulations. Simulations involve the viewer being immersed as a participant to some degree, however, instructional videos engage the viewer visually while he or she performs certain tasks in conjunction with, or subsequent to, viewing the video. In this chapter, I apply attributes of the model to instructional video, specifically machinima video. Machinima video is a video product that integrates virtual environments, much like the environments described in the previous chapter. These videos are not as interactive as simulators like those used in the previous chapter, however, they are used for instructional purposes. The model's principles can be applied to analysis of these much as they were in the previous chapter. I illustrate these applications in this chapter to suggest further study of these kinds of instructional tools with the model.

Simulators like SL engage visual, aural, spatial, and behavioral modes of representation in an immersive multisensory experience similar to actually performing the task in the real world. Further, demonstrations in SL can be recorded like videos and shown subsequently. This allows viewers to see particular movements and motions associated with a task before they actually attempt it. Consequently, such videos can be analyzed in terms of intermodal sensory redundancy, temporal synchronicity, and attention-modal filtering. In particular, such video can be studied relative to its rhetoric and sensory experiences applying new theories of multimodal assessment. Many of these use concepts integrated into the model. I apply the model to two case studies of machinima video instructional products. One was developed by a student and another developed professionally. I have presented information about these products previously (Remley, 2010, 2014). However, as with the cases presented in the previous two chapters, that discussion was framed within different theoretical foci. Nevertheless, they can be used to illustrate attributes of the model presented here.

MACHINIMA INSTRUCTIONAL VIDEO 1: SECOND LIFE ORIENTATION

The use of 3-D virtual environments in education and industry has grown considerably in the past 10 years. Stephanie Vie (2008) stated that, "*Second Life* has grown in popularity among educators because of their interest in the possible pedagogical uses of video and computer games, including their potential ability to strengthen students' critical media literacies." Also, Dickey (2005) reported on affordances and constraints of using *Active Worlds*, another 3-D virtual environment, in education. Such technologies have also appeared in workplace settings, as I observed previously (Remley, 2010). However, illustrating the need for critical media literacy, Malaby (2007) observed that some companies that have experimented with *Second Life* are leaving it after ascertaining limitations in its ability to help them operate. Nevertheless, companies continue to use *Second Life* and similar 3-D virtual world environments.

Video has become such a prominent tool that instruction in video composition has become part of pedagogy in rhetoric (Eastern Washington University, 2008; Jones, 2007; Selfe, 2004). Many instructional videos can be found through YouTube, and these videos vary from real-life video products to machinima video products such as those generated within virtual environments like *Second Life.* Lowood (2005) defined machinima as "making animated movies in real-time with the software that is used to develop and play computer games" (p. 10). While *Second Life* is not itself a game—there is no game theory built into its system, and Linden Labs, the environment's owner, refutes any attempt to declare it a game (Kalning, 2007)—its interface looks much like that of a video game.

Much as video game sales have proliferated with new 3-D gaming systems, 3-D virtual platforms are being used for business and industrial uses, ranging from facilitating meetings, presentations, and training that includes video training (Wagner, 2007). For example, the pharmaceutical company Exergen, a maker of digital heat-sensitive thermometers, uses machinima to show how its thermometer works. Also, over 20 companies produce machinima videos (Linden Labs, 2010), suggesting demand for machinima video design skills. These videos integrate several of the principles of the model, and they suggest that an understanding of these principles can facilitate production of effective videos.

A student of mine developed an instructional video providing an orientation to the 3-D virtual environment *Second Life*. He developed the video in response to a term project for a business writing course in which students were invited to develop a multimodal product. I have reported on assessment of this product (Remley, 2012), but I include related information here to reinforce attributes of video that affect cognition relative to the model.

Odell and Katz (2009) identified some criteria related to rhetoric and reader perspective that can be useful in assessing multimedia products; some of these have a direct relationship to principles of the model. They called attention to audience-related attributes such as using the audience's prior knowledge to facilitate new information (prior experience); creating and meeting an audience's expectations to understand a given message or point (intermodal sensory redundancy and attention-modal filtering); selecting and encoding images toward meeting cultural expectations or perceptions (prior experience and intermodal sensory redundancy); and logical/perceptual relationships that facilitate understanding the relationships between developmental points of a given message (intermodal sensory redundancy, attention-modal filtering, and temporal synchronicity).

As I (Remley, 2012) note, video production and design attributes such as camera angle and image-audio narration combinations can affect the ability to meet an audience's information needs. For example, a camera angle that integrates irrelevant information, such as shown in Figures 7.1 and 7.2, will impact attention-modal filtering. In the related video, instructions on how to perform certain tasks in SL with a Mac machine are provided. In the portion of the video from which the particular screenshots are taken, the instructions revolve around how to use camera control tools to change the position of the camera, however, the focus is on moving the camera from side to side or up and down, not on establishing a given distance from the avatar's position. There is far too much space between the avatar's head and the top of the screen. A viewer needs to filter out the sky and background in order to focus on the action described.

Figure 7.1. Too wide of a camera angle.

Figure 7.2. Too much space above head.

While the viewer observes the action on this screen, the audio narration occurs simultaneously, explaining the narrator's actions to manipulate the camera positioning. This integrates two modes—visual and audio—that help to explain the action. Echoing Baddeley's (1986) and Moreno and Mayer's (2000) findings to limit the number of modes to two, it helps focus attention while engaging multimodal neurons. Again, this limits the amount of filtering required.

However, the information is not necessarily redundant. The visual information shows what the action looks like, while the narration explains how to perform it. This may create a challenge for the viewer relative to helping facilitate cognition. Two modes are integrated, but the information related to each is not closely linked.

So this product has some good attributes relative to the model, however, I have also identified a few weaknesses that could be addressed by applying elements of the model better. The next section details analysis of a professionally developed video pertaining to soccer skills.

SOCCER VIDEO PRODUCT

Challenger Sports Corporation produces instructional videos to help children and teens learn various soccer skills and advanced moves. The video includes a series of short instructional machinima video in which a virtual boy or girl demonstrates a given skill. Each segment is set up applying a number of elements of what I described in the chapter on Training Within Industry as well as what I mentioned in the previous chapter about other new media applications and machinima video.

Each segment begins with a narrator introducing the particular skill that will be demonstrated. The male voice identifies the name of the skill by its technical title and when it likely would be applied or used. After that introduction, the boy or girl demonstrates the entire skill in regular motion. As with TWI, one shows the audience the entire operation.

After that demonstration is completed, the boy or girl stops, and the narrator presents a step-by-step listing of what actions the skill or move involves. As the narrator identifies a given step or action, the step appears in text to the side of the image of the boy or girl. When the narrator finishes describing the step, the boy or girl performs that step and then stops to wait for the next step to be listed and described. The same sequence is repeated until all steps have been presented and demonstrated. This is similar to the portion of TWI training wherein the trainer breaks down the task and talks through the important attributes of each step.

Next, the narrator introduces a slow motion version of the demonstration, and that is followed by the demonstration in regular motion. Again, in TWI's program, one would slowly go through the steps and then let the trainee perform the steps. The slow motion demonstration allows the viewer to observe more closely the action with each aspect of the skill.

A number of neuro-friendly elements are associated with the rhetoric built into this kind of presentation and demonstration. I analyze this video as another example of a case study that applies the integrated model.

Neurorhetoric

I call attention to three specific items of the model: intermodal redundancy, attention-modal filtering through temporal synchronicity, and prior experience. These integrate biological as well as social phenomena within cognition.

Intermodal Redundancy

As indicated above, each segment integrates visual and audio modes. Generally, the audio narration explains what is taking place with the visual demonstration. The visual demonstration shows the viewer what the particular skill or step looks like. This allows the viewer to learn what is involved with a given skill or step before attempting it themselves.

While the viewer cannot touch the ball as the boy or girl does, he or she can observe the degree to which the boy or girl is striking the ball with his or her foot. This helps the viewer anticipate how to hit the ball when he or she attempts it him or herself.

Temporal Synchronicity and Attention-Modal Filtering

While the demonstration includes both audio and video, the narration and visual demonstration rarely occur at the same time. In Moreno and Mayer's (2000) experiments, generally, they tested cognition when audio occurred at the same time as the visual image, or text was placed with an image at or about the same time as the image appeared; and they tested cognition when the two modes were separated briefly or presented concurrently rather than simultaneously. They found that, generally, cognition is better facilitated when the two modes are presented consecutively and when verbal information is presented in audio form rather than in print. They labeled these phenomena the "temporal congruity principle."

In the example of the machinima video in the previous chapter, the audio narration and video action occurred simultaneously, which could cause challenges for cognition depending on how much information is involved. Again, Moreno and Mayer (2000) and Baddeley (1986) encouraged limiting the number of modes involved to two to optimize short-term memory and recall. However, splitting them temporally, even for a second, helps to address some of the challenges of the Colavita visual dominance effect.

Addressing the Colavita Visual Dominance Effect

According to the Colavita visual dominance effect, when visual stimuli are accompanied by audio stimuli, one perceives that the visual preceded the audio, and more attention is paid to the visual. The audio acts, consequently, to help explain the visual, but the brain has to manage information coming from the two different stimuli. This is not a large challenge, since the brain has multimodal neurons to assist this processing. However, if the two stimuli are clearly separated but placed closely to each other in sequence, they can enhance the focus on a given set of information. The temporal synchronicity facilitates intermodal filtering while acting to explain or facilitate a visual representation of each step.

Mirror Neurons and Prior Experience

As mentioned before, mirror neurons allow an observer to experience vicariously the movements he or she is observing in another. The mirror neurons act to process the specific actions associated with the task. If one has played soccer to any degree—even just dribbling a ball across a room—before viewing the video, one will be able to process how hard to strike the ball with each skill or step as he or she views the demonstration. If one never has dribbled a soccer ball, it may be more difficult to gauge how hard the demonstrator is hitting the ball.

Through prior experience, then, one comes to understand how a soccer ball reacts to various touches—a tap, a push, or a soft strike (short pass) or hard strike (long pass or shot on goal). This experience enables the mirror neurons to better process the information related to touching the ball with the foot.

CASE STUDY OF SKILL SEGMENT

A particular skill demonstrated in the video is called the "inside turn." A player who is dribbling the ball and who wishes to change directions quickly has to maneuver him or herself as well as the ball. The faster the player can perform the turn the better he or she will be able to outmaneuver an opponent. The particular segment of the video related to this move teaches the player to turn 180 degrees—a full turn to the opposite direction.

The narrator introduces the segment, acknowledging the particular skill name and when it may be utilized. The narrator then introduces the video demonstration. Then the player-avatar demonstrates the skill in regular motion (see Figure 7.3) without narration. This enables the viewer to see the entire process without interruption while focusing attention on the visual mode.

Figure 7.3. Initial demonstration: full speed.
All virtual soccer images provided by and used with permission of Challenger Sports Corporation.

The viewer observes the entire process and is able to see what it all looks like. This provides a visual representation that is comprehensive in the information it presents.

After the regular speed demonstration, the narrator introduces the various steps involved in the process. First the narrator acknowledges the step, and the step is shown verbally with text on the screen. This integrates two modes simultaneously—visual (print-linguistic text) and audio (narration). Then the avatar demonstrates that part of the process at regular speed with the textual information still on the screen. When the step has been demonstrated, the narrator then acknowledges the next step and the verbal information appears on the screen followed by the demonstration of the step (see Figure 7.4).

Presenting the verbal information textually and through audio narration facilitates reinforcement of verbal information through visual and auditory modes.

Once the entire process related to the skill is broken down and shown, all the steps remain on the screen (Figure 7.5). This reinforces the visual information shown in the initial demonstration verbally.

Finally, the entire process related to the skill is demonstrated visually at half speed so the viewer can watch each part of the process more closely, but in connection with the entire process without interruption (Figure 7.6). The verbal information remains on the screen, acting as a sort of checklist as the viewer observes the demonstration.

This view, again, limits the attention to the visual, but facilitates closer viewing of it so as to encourage better cognition. Doing so facilitates attention-modal filtering, as the viewer can look at both the demonstration and the verbal information.

Finally, the entire process related to the skill is shown again at regular speed without audio narration or verbal information presented textually (Figure 7.7). This enables the viewer to observe the entire process as it ought to be performed, however, the information is better understood after the verbal explanation is provided in multiple ways.

Throughout the segment, the visual and verbal information is separated so as not to force the viewer to process the different information at the same time. Further, the verbal information is provided two ways (textually and audio narration) to limit short-term memory dynamics to those two modes—visual (print-linguistic textual information) and audio. Providing it both ways also helps to reinforce that information. Ultimately, then, the viewer experiences the instructional information three ways: visual demonstration of the avatar, visually through the textual information, and aurally through the audio narration.

Discussion

With respect to the intermodal redundancy, temporal synchronicity, attention-modal filtering, and prior experience principles, the segment exhibits a balance

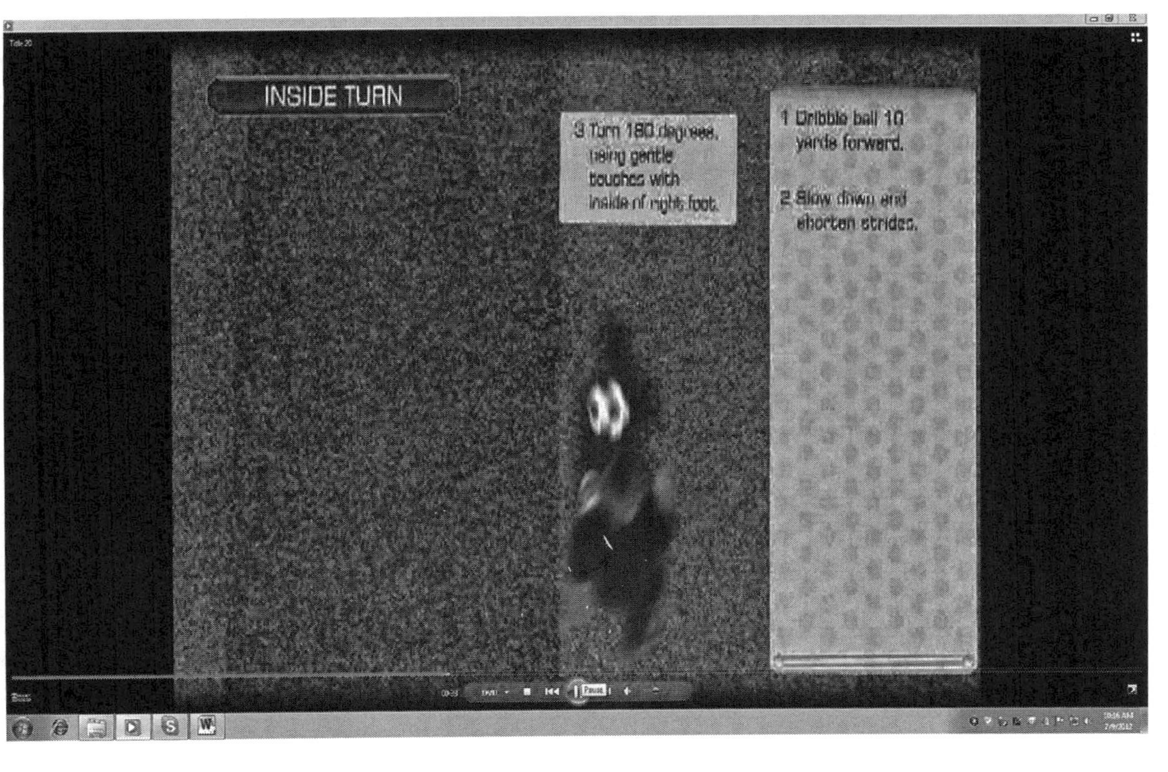

Figure 7.4. Breaking down step: Full speed, but with stops.

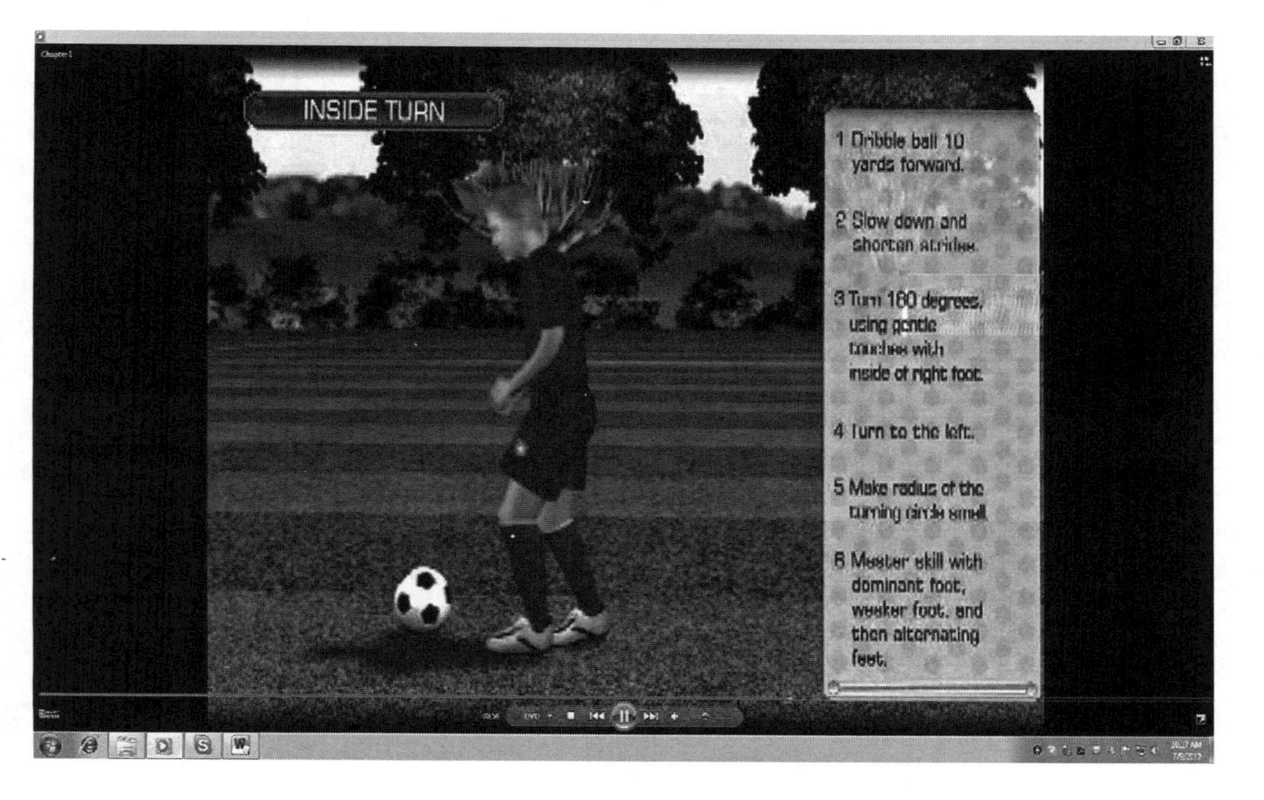

Figure 7.5. All steps shown in box.

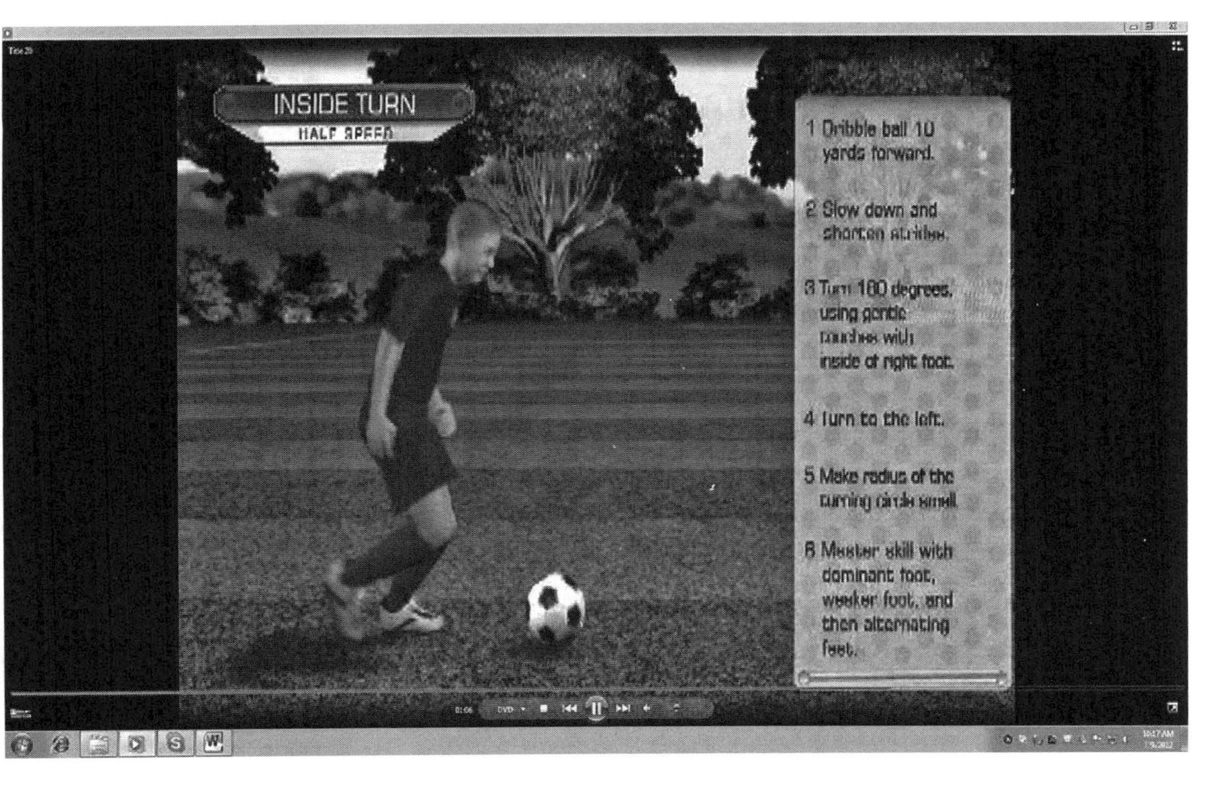

Figure 7.6. Full demonstration at half-speed.

Figure 7.7. Full demonstration at regular speed.

between various modes, implementing the modes consecutively (for the most part) and facilitates some cognition about touch through the viewer's prior experiences. Again, two specific modes are used—visual and audio—and they are presented at different times.

Information about the skill is shown visually through the demonstration at full speed (twice) and at half speed. The viewer thus sees the entire process performed no fewer than three times. Further, information is separated temporally to limit attention required. Steps are itemized and shown one-by-one visually and described in the audio narration. This facilitates cognition by focusing attention on each step in the process while integrating multiple modes to reinforce the information. This permits the multimodal neurons to act toward integrating redundant information while balancing the limitation of the number of different modes and senses being used to process the information.

Finally, if the viewer has played soccer before, he or she can gauge the amount of effort used by the avatar to strike the ball as he or she works through the process. Mirror neurons developed through prior experience will help the viewer understand that effort. Mirror neurons facilitate an understanding of visual and motor actions.

However, some limitations exist. For example, analysis of SL simulation would likely exclude self-motion analysis, which is popular in neurobiological scholarship. One moves the avatar via a mouse, so immersion is limited. Also, because it is a heavily visual experience while limiting other sensory experiences such as touch or smell, analysis of the visual dominance attribute may be limited to understanding the impact certain visual information has on perception and cognition. This can also facilitate some analysis of attention-modal filtering.

Again, study is limited by the new media technology's capabilities, and this can be considered separately within the study of technologies. An understanding of a technology's affordances and constraints can inform how educators/trainers use a given tool for training purposes.

CONCLUSION

As with the applications with hands-on training methods, multimodal print materials, slide shows, and simulations, the model can be applied to facilitate analysis and development of video products. Because of differences in the ways viewers interact with videos compared to simulators, the principles of the model are applied a bit differently. Nevertheless, they are still applicable. While the video products discussed in this chapter were machinima video, rhetorical attributes and biological attributes of are easily applied to "real life" video products, since many of the same rhetorical and biological attributes apply.

CHAPTER 8

Comparative Neurorhetorical Analyses

In the three previous chapters, I have detailed neurorhetorical analyses relative to particular media and using different topics of instruction in each case. In this chapter, I provide analyses of the same instructional topic presented using different media to show a comparative neurorhetorical analysis. That is, how do neurorhetorics play out relative to particular media when the topic and audience are the same? For these analyses, I have chosen the topic of web-design tutorials, specifically, those featuring Adobe Photoshop as the design tool. Instruction in web design is often included in technical communication coursework. However, how that instruction is delivered can vary with the various media available. The various media involved in the discussion in this chapter are print-linguistic instructions that include graphics, presented within a pdf file; a PowerPoint slide show; and a video that includes narration.

The subjects of the analyses were found from searches conducted on the Internet on the topic of "web design, tutorials." I refined that search term to include "video" or "PowerPoint" for selecting the particular products. Locating a PowerPoint slide show that included audio narration was challenging. None appeared in the first five pages of the search-result listing for the term that included "PowerPoint," however, the one I chose to use appears not to be associated with a real-time presentation. Thus, we can assume that there would not be narration (though not digital) accompanying the slide show. I have received permission to use the products and images from the producer/author of each product.

The analysis focuses on specific steps related to web design using Photoshop, and I take particular information—textual and graphic—from each product and consider how the particular medium facilitates neurorhetoric associated with those steps. Throughout, I compare and contrast the neurorhetoric relative to the affordances and constraints of the medium. As with the other chapters, the analysis focuses on applying the terms identified in Chapter 3 to a product.

However, experimental design research can examine specific neural activity associated with such products and assess their effectiveness in cognition.

CREATING A WEBSITE USING ADOBE PHOTOSHOP

For this analysis, I focus on the initial steps of creating a website using Adobe Photoshop CS3. These steps involve creating the first two layers of the site: the general background and the foreground layers of the banner. Because of the different media involved, I call attention to how the media affect the design, contributing to affecting the neurorhetoric. One attribute of the model that seems disregarded in the analysis is the prior experience attribute. In each case, the designer seems to build into their instructions the assumption that the viewer has little to no prior experience with web design or with Adobe Photoshop. This is actually a characteristic that enables this analysis; the same assumption is associated with the various designs, minimizing the effect prior experience may have in the analysis.

Print-Linguistic Text: Alphanumeric Text-Image Design (HTML/PDF)

Carlos Aleman designed a set of instructions and posted these to his website. Figure 8.1 shows screenshots scrolling the entire page that listed the sequence of steps and related images that appear in this set of instructions.

Set in a step-by-step format, an image of the screen is placed with each step to show the reader what the screen looks like as the reader performs a given step and where to locate particular tools on the screen. This is generally considered good instructional design since textual information is presented alongside visual information. The text tells the reader what to do while the image shows what the action looks like on the screen.

The design recognizes the intermodal sensory redundancy principle in that, while all information is presented in print, some is alphanumeric, and there are also images. This also facilitates the visual dominance attribute, allowing the reader to view the image without interference of other modal stimuli. Because visuals appear with each step and the figure is related directly to the step, modal filtering is facilitated too. That is, there is not any extra, irrelevant information that one must filter from the information.

These instructions also seem not to assume a prior experience with web-design tools; Aleman includes information about what codes to use for particular color attributes while also acknowledging specific keystroke actions.

This design works better for an audience that has less background with web design; though it assumes one can use a keyboard to locate certain tools in the menu. Movement about the screen is not provided since the images are static,

Free Web Design Tutorial

Creating an HTML/CSS Web Page using Adobe Photoshop and Dreamweaver

Part 1

Many people have told me that they wish that they could design web sites. So I thought I'd create this tutorial, since there doesn't seem to be too many free courses online that teach both the graphic design skills and HTML and CSS necessary for creating web pages.

I would like to *outline* (not a good word for right brained people like me) a *workflow* (?) for creating a web page using my home page as an example and Windows as the operating system (Mac users will have to find alternate keyboard shortcuts). Since I am a self taught designer, I'm not sure how unconventional my methods will seem, but hope this will be helpful. As you know, technology quickly becomes obsolete, so may the expiration date on this project live long and prosper...

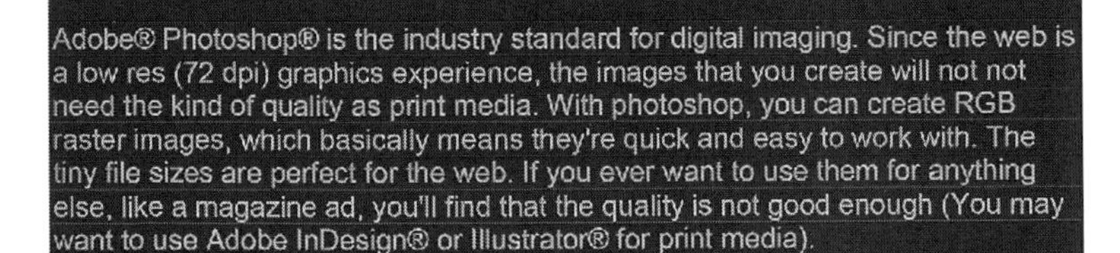

Adobe® Photoshop® is the industry standard for digital imaging. Since the web is a low res (72 dpi) graphics experience, the images that you create will not not need the kind of quality as print media. With photoshop, you can create RGB raster images, which basically means they're quick and easy to work with. The tiny file sizes are perfect for the web. If you ever want to use them for anything else, like a magazine ad, you'll find that the quality is not good enough (You may want to use Adobe InDesign® or Illustrator® for print media).

The types of files we will use are jpgs and gifs. Png files are the best, but we'll have to wait until everyone stops using old web browsers. I use png files all the time with Flash projects (See my Flash Tutorial) because of the transparent effect in animations.

Figure 8.1. Entire webpage.
Used with permission of Carlos Aleman.

I don't know which version of Photoshop you have, so you'll have to at least play around with it for a few hours to find out how the layers link together and other basics.

First, I'll start off with a blank canvas. Because computer screen resolutions are becoming increasingly wide, very few people still have their displays set to 700x600 to surf the web. I'll start with 1000x700 pixels, because that is a comfortable size for me to work within Photoshop and it will allow me to create a site that is a good size.

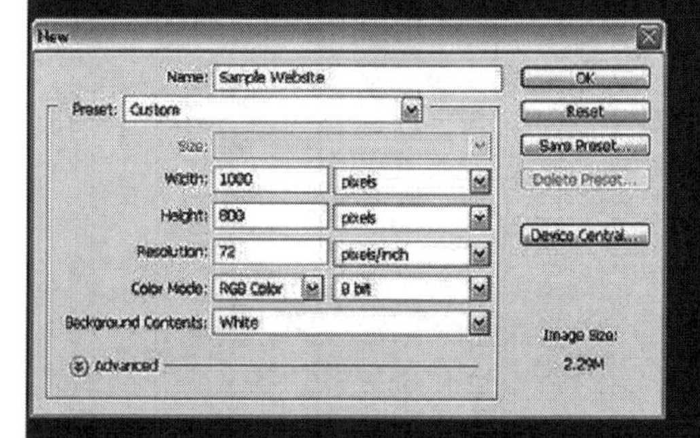

Select black (#000000) in your foreground color and then hold down Alt>Delete to fill the canvas with black as your background color. You can create gradient and textured backgrounds that tile, but here we're just using solid black.

Create a new layer. Make sure that your rulers are set (View menu>Rulers). With your selection tool create a rectangle that is 770x190 pixels. You can tell the height and width by looking at the Info palette in the windows menu as you're creating the rectangle. Select a color in your foreground color picker and then hold down Alt>Delete to fill the rectangle with color. I used orange for visibility. Drag and drop vertical guide lines from the ruler area to the dotted lines of the selection rectangle. Hold Down Control>Delete to deselect the rectangle.

Figure 8.1. (Cont'd.).

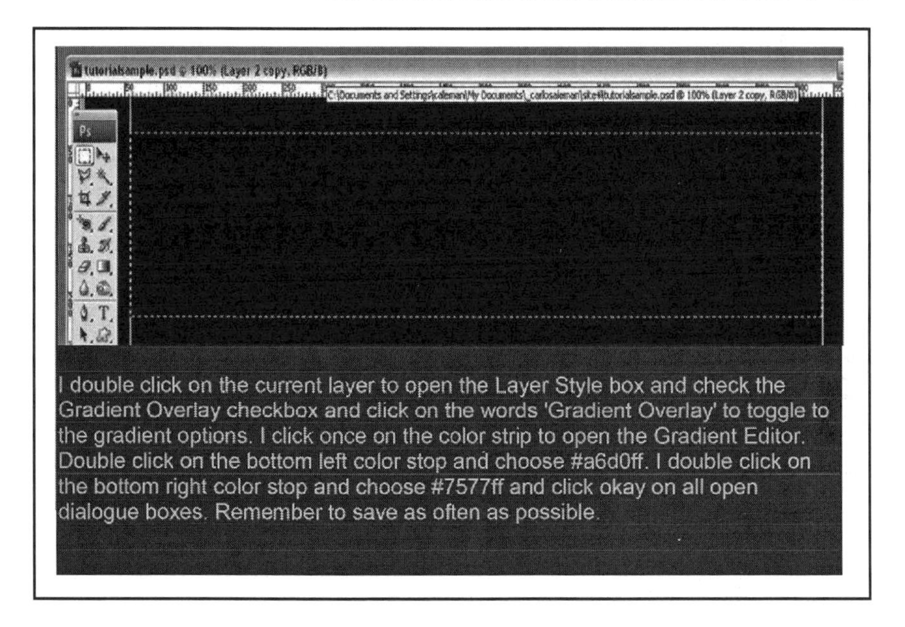

Figure 8.1. (Cont'd.).

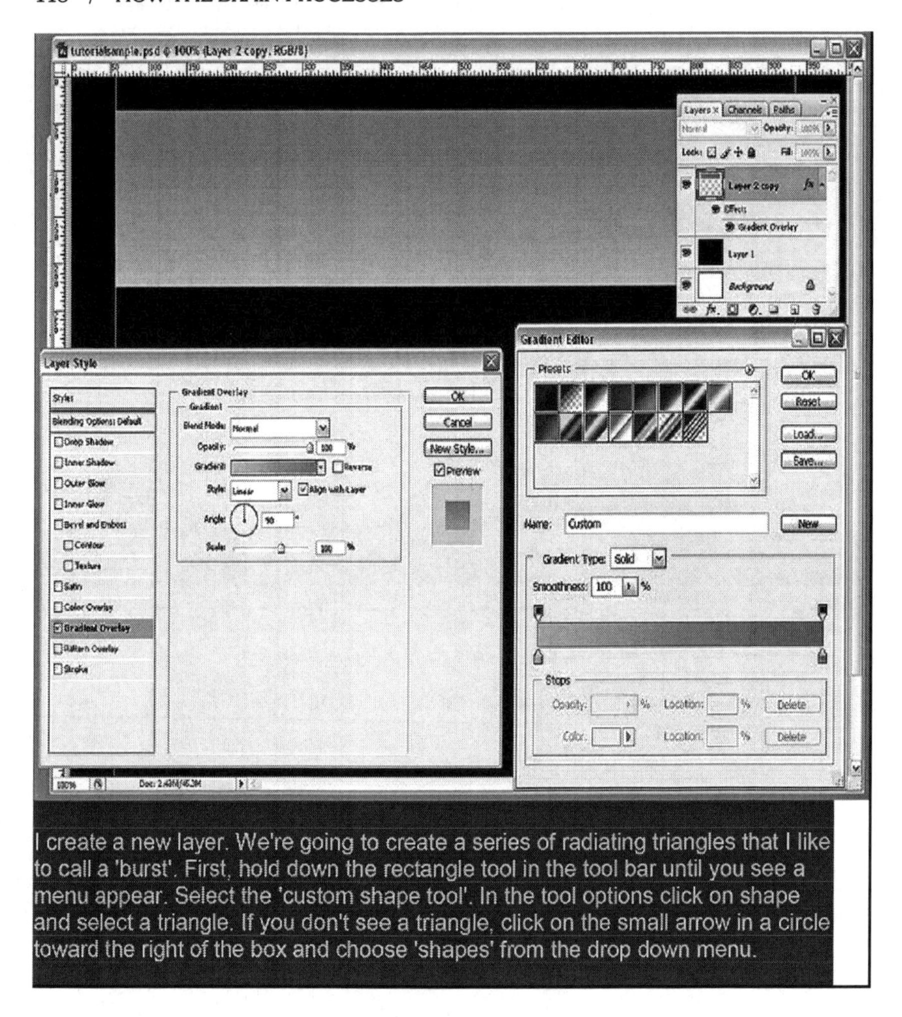

Figure 8.1. (Cont'd.).

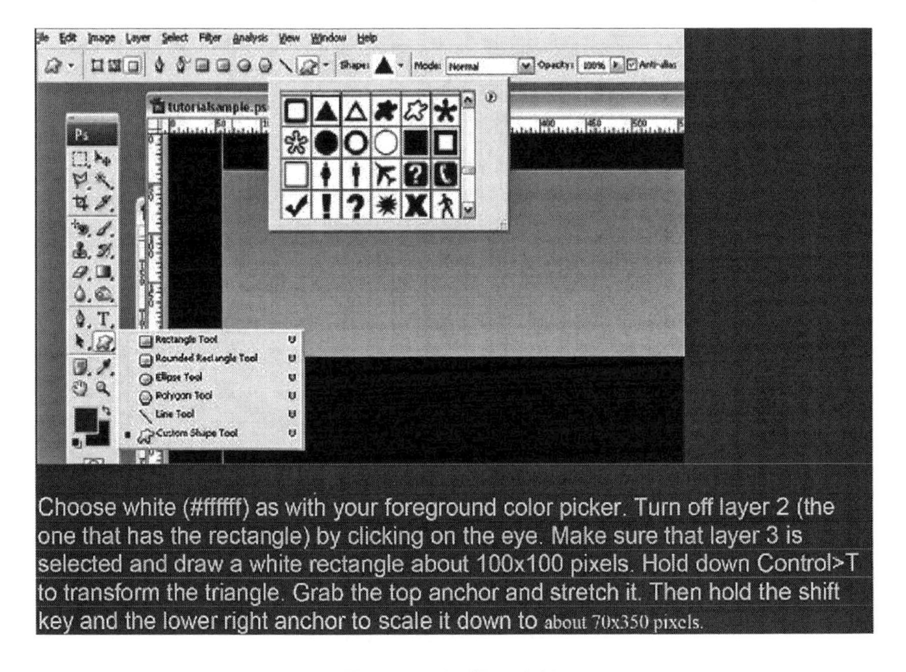

Figure 8.1. (Cont'd.).

limiting the potential for mirror neurons to acquire information. Video that includes screen capture of such movements can permit acquisition of such information for mirror neurons, as indicated with the next example.

Video

Ed Johnson developed a video that he posted on YouTube. The video lasts a little over 12 minutes, and the particular portion associated with the steps I provide in this chapter takes about 2½ minutes. Figures 8.2 through 8.6 show screenshots of a sampling of specific parts of that portion of the video.

Of course, Mr. Johnson narrates as he moves through each step, the video portion capturing the action on the screen. Such video enables the viewer to experience the visual dominance attribute of the model while also experiencing temporal modal synchronicity with the audio narration associated with the visual information provided at the same time. While this facilitates intermodal sensory redundancy, this goes somewhat against the modal-filtering principle as well as the visual dominance attribute. As Moreno and Mayer (2000) found, information presented through two modal stimuli at the same time may not be as effectively understood as information presented in sequence, as with the machinima video described in the previous chapter.

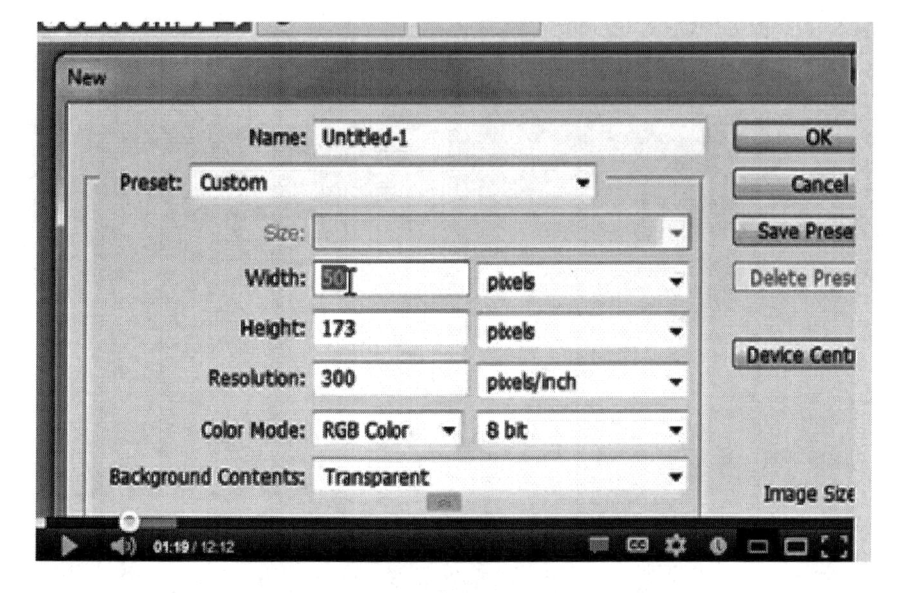

Figure 8.2. Initiating process.
All images from video used with permission of Ed Johnson.

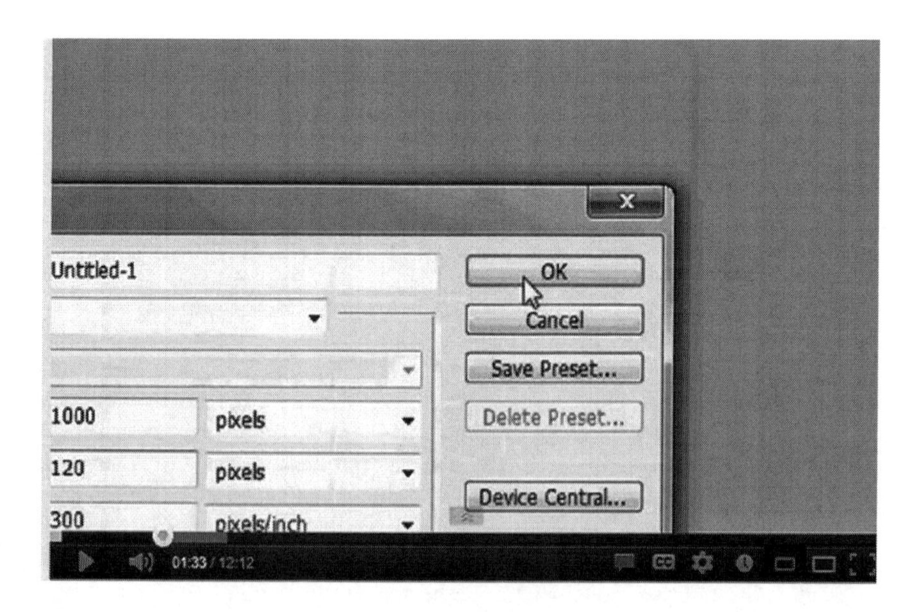

Figure 8.3. Step 2.

Figure 8.4. Another step.

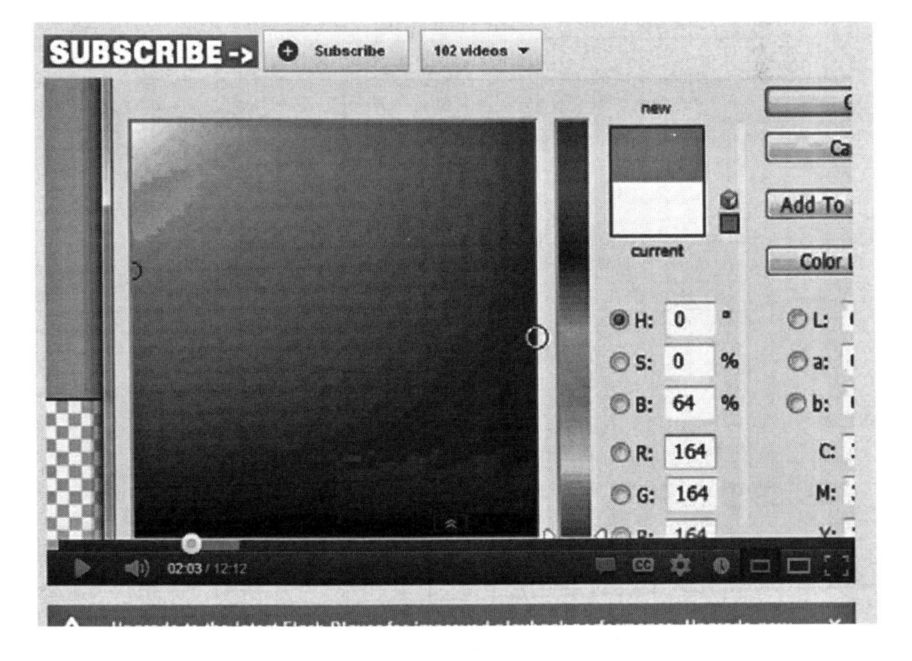

Figure 8.5.

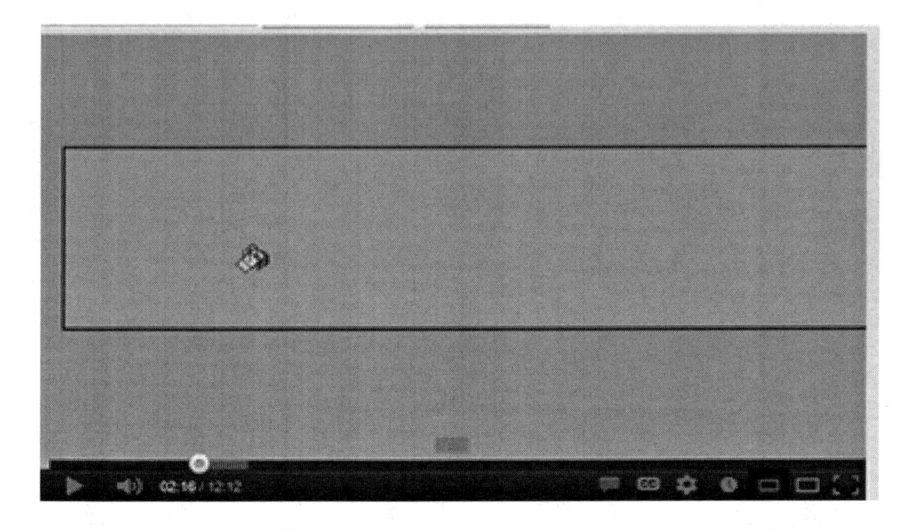

Figure 8.6.

However, the screen capture ability of the medium facilitates acquisition of information by mirror neurons that will allow the viewer to duplicate the movements with the cursor and locate the particular information easily. The ability of the medium to include narration also allows for the aural neurons to be involved in processing information. The narration is a bit fast however, and this may negatively affect learning. However, one is able to review the portions of the video wherein that narration is important.

Print-Linguistic Text: PowerPoint and Interactive Narration

Lauren Thompson developed a PowerPoint slide show in which she explains the process. Each slide describes attributes without explaining what to do or showing specific images from the screen. Figures 8.7 through 8.11 are a sampling of the slides associated with the steps involved in creating the initial attributes of the web page. While the slide show is available publicly on the Web, it does not include narration with it, yet it is important to understand that she developed the slide show to be used in conjunction with real-time narration and interactive instruction. This raises an issue regarding viewer experience of the slide show: Is one experiencing it with the workshop narration and instruction, or is one experiencing it as a stand-alone slide show? The difference affects the neurological processing experience too.

The slide show was originally designed to be used within an hour-long workshop setting that also included a demonstration screen, and attendees had access

Getting Started

File > New...

Select the size of the blank document that you want to create. Pay special attention to your choice of resolution. If you're creating an image for the web, 72 pixels per inch will suffice. If you're creating a detailed art project, you want at least 300 pixels per inch, but obviously the more the bettter.

Figure 8.7. Initiating process.
All slides shown with permission from Dr. Lola Thompson.

A Few Words About Color

Computers recognized many different color formats. Some common choices include:

1) RGB color - lists a value between 0 and 255 for each of the {Red, Green, Blue} components of a color. (0, 0, 0), the absence of color, is completely black. (255, 255, 255), the presence of all colors, is completely white. Most Photoshop artists work in RGB.

Figure 8.8. Caution.
All slides shown with permission from Dr. Lola Thompson.

A Few Words About Color

2) CMYK color - lists a percentage of each of the colors {Cyan, Maroon, Yellow, blacK}. CMYK can best be thought of as the "color printer cartridge colors." Some graphics designers choose to convert their work (after it has been done in RGB) to CMYK so that they have a better sense of how it's going to look after it is printed.

Figure 8.9. Caution 2.
All slides shown with permission from Dr. Lola Thompson.

Instructions

1) Create a black background
2) Write text (should automatically open in a new layer)
3) Hide background layer so that you can see the text
4) Make a Duplicate Layer. Change the color of the text in this layer to white.
5) Select Gaussian Blur for the Duplicate Layer from the Filter Menu.
6) Move the Duplicate Layer beneath the Text Layer on the layers list. Now feel free to move the Duplicate layer in relation to the Text layer in order to get the drop shadow effect that you want.

Figure 8.10. Steps.
All slides shown with permission from Dr. Lola Thompson.

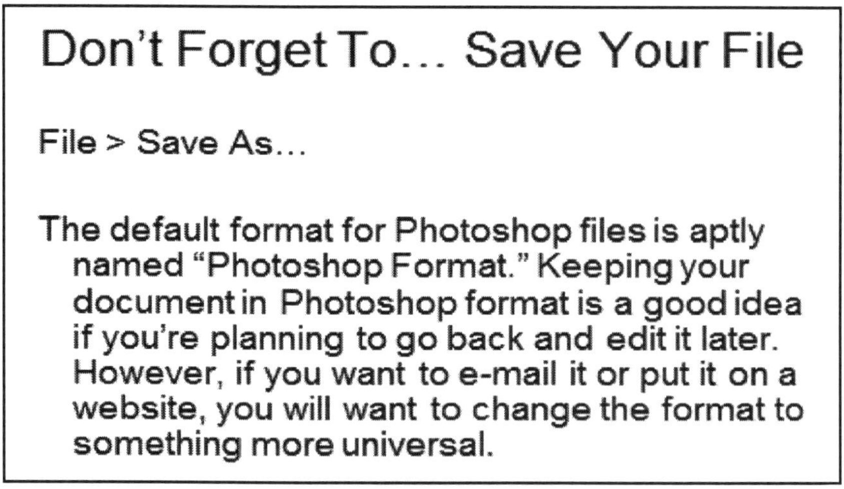

Don't Forget To... Save Your File

File > Save As...

The default format for Photoshop files is aptly named "Photoshop Format." Keeping your document in Photoshop format is a good idea if you're planning to go back and edit it later. However, if you want to e-mail it or put it on a website, you will want to change the format to something more universal.

Figure 8.11. Saving file.

to a computer. Dr. Thompson explained that, "Participants sat at computers while I demonstrated techniques on an overhead projector (I would alternate between the slideshow and a Photoshop window). After each demonstration I would give the participants a few minutes to try the techniques for themselves while I walked around to offer assistance" (L. Thompson, personal communication). So there were other stimuli used in conjunction with this slide show—visual and audio stimuli, and also spatial and behavioral. However, it is publicly available online outside of that initial setting and can be used instructionally by oneself or with narration from someone else.

Generally, all the information is presented in alphanumeric print-linguistic text. While a step-by-step approach is used here, the slide show seems to describe the process more than explain the steps involved. Thompson provides background information about the particular attributes of creating the page before getting to the steps associated with the actual process of creating the page. Narration used in conjunction with the slides likely talked workshop attendees through the processes described in the slides. This is limited, though, to size and color attributes. This allows one to learn these attributes separately from the process, facilitating filtering of information.

An assumption the slide show seems to make is that the viewer has some prior experience with web design in that it does not provide any information about where to locate tools associated with size and color. However, this information may have been included in the narration and facilitated with the demonstration screen in the workshop setting for which it was originally designed. Also, modal filtering is involved in this case in the sense that visual images are omitted—only print-linguistic text is provided. So the viewer would see only textual information on the slide show screen while being able to move to the visual image of the screen on the demonstration screen, however, this act integrates modal filtering by separating the modal information onto different screens, requiring one to view each separately.

Because audio narration does not accompany the slide show as it appears online for public viewing, though, one can move through it at their own pace. One can even skip through the preliminary slides to get to the instructions if they have prior experience with those steps in another design tool. Further, the instructions seem to assume prior experience with web-design tools. So there is some attribute of prior experience involved if one uses it outside of that initial workshop. This is okay, though, per the observation by Amerine and Bilmes (1990) that any set of instructions builds into it certain assumptions about the audience's familiarity with related concepts and tasks. However, a user would need to understand this limitation as he or she began using the tutorial.

While this set of instructions includes no graphics, it may be effective for an audience outside of the workshop setting that has background with web design and wants more information to learn how to use colors more effectively in designing a web page. In a workshop setting, one may be able to receive additional instruction regarding web-design basics separate from the slide show.

LIMITATIONS OF MEDIUM

The three products presented in this chapter are designed using different media and are associated with three different viewing experiences. As such, the medium affects to a large degree what one can design modally into the product. A print-linguistic document like that of the Aleman set of instructions allows for visuals to be included and textual information to be integrated with each step. There is modal redundancy here. A PowerPoint slide show facilitates narration, yet the example included here assumes no such narration occurred.

However, the video product facilitates the display of cursor movement about the screen, facilitating mirror neurons. Further, the inclusion of narration in the video allows for some additional information not provided with just the video, helping to explain steps further. This allows for the visual dominance attribute as well as the intermodal sensory redundancy principle, though Johnson's narration is not synchronized in succession with the visual information. This negatively impacts redundancy and cognition.

Finally, the slide show that would be used in conjunction with other modes in a workshop facilitates additional stimuli, and the separation of textual information from the other visual information acts as a form of modal redundancy for narration—audio information is presented visually through the print-linguistic text, reinforcing it. The viewer also needs to shift their vision to a different screen to see the actual interface, allowing some attributes of the Colavita visual dominance effect to play out.

As I pointed out though, because the slide show is also available outside of that workshop setting, it takes on different attributes as an instructional product. The other forms of representation used in the workshop are not available in that case, and the viewer needs to have more prior experience with web design in order for the slide show information to be useful.

CONCLUSION

I provide this comparative analysis because of the importance of understanding how a medium affects modal capacities in design decisions and consequently, cognition. Kress and van Leeuwen (2003) have discussed limitations and capabilities of media on design and modal attributes therein, however, it is important to consider how decisions to use certain media affect neural processing. Any decision by a designer of multimodal instructional materials has to include consideration of the technologies' abilities and limitations.

With regard to a PowerPoint slide show, for example, one expects there to be some narration, either digital or in real-time from the presenter. A slide show presented as a stand-alone tool without narration assumes that the viewer will be able to fill in some blanks about the information and how it is to be "read" or interpreted. However, a website that includes text and graphic images can better convey the instructional message.

CHAPTER 9

Research Implications

I have described a variety of applications of the new model relative to the kinds of instructional materials that one may compose using different new media. Those chapters were limited to content analyses of specific multimodal/multimedia instructional or training materials and based on what may be considered "normal" neural processing of those stimuli. In this chapter, I describe a variety of ways that the model can contribute to further research in education and workplace training, possibly affecting instructional and training materials. Given the inclusive nature of the model, that is, including biological sciences aspects of neuroscience and social science attributes of it, the model facilitates several kinds of research. Because it draws on scholarship and theorization from both biological sciences and social sciences, it facilitates the kinds of research that have already been conducted in each area but reframes the analyses of them. It also opens new doors to research that may include both biological as well as social dynamics of cognition. Because it is a descriptive model, rather than prescriptive, particular attributes of it may be analyzed independently of other attributes, however, theorization needs to value that the synthesis of the various attributes contribute to cognition.

As I describe the ways the model may contribute to research and design of new instructional and training materials, I recall some suggestions I offered in previous chapters. As I detailed the new model, I offered ideas for possible research relative to commonly used approaches in the rhetoric and neurobiology fields, and as I described potential applications, I also suggested certain kinds of research. I elaborate on those application suggestions here.

The model facilitates not just research into cognition, but it also facilitates design of productive materials to optimize cognition. Instructional technologists, technical communicators, and cognitive psychologists are all very much interested in understanding how design of instructional materials affects cognition. Such materials are important to education and training. Consequently, the research indicated above should be multidisciplinary in nature. Researchers from the different fields—social sciences, humanities, and physical sciences—doing joint projects will benefit from each others' understanding of research designs used most in each field.

EDUCATION

An important implication associated with reward neurons and education is that instructional materials should include statements to help students understand the value of a given topic associated with the materials. Such information, even a short paragraph or bullet point accompanied with narration to re-enforce it, may prompt reward neurons to release chemicals that will motivate students to want to learn the information.

Some services that assess the effectiveness of instructional materials in Web-based settings, such as Quality Matters, include this consideration in their criteria. They encourage explicit statements in materials to help students understand how particular information and activities fit in the course's learning objectives so students understand the connection between what they are being taught and what skills the course is trying to develop. Again, an explicit statement acknowledging a reward associated with the lesson will help the student become motivated to pay attention to it. Relative to the model, this information should be conveyed visually and orally to re-enforce each other.

Multimodal Instructional Materials

The more new media technologies evolve, the more modes they can use and the more stimuli they facilitate. As suggested with the intermodal sensory redundancy principle, the brain likes to drawn information from various modes to facilitate optimal cognitive processes. So media tools that engage a number of senses tend to be encouraged in instructional materials.

While various media can be integrated into classroom pedagogies, two particular ways this model informs development of such tools is relative to valued modes of instruction—web-based education and service-learning/internships. Both of these integrate several modes into instruction, and research can consider optimal combinations of stimuli relative to each approach.

Web-Design/Web-Based Instruction

As institutions attempt to facilitate more convenience for students to get their education, more courses are being offered via web-based delivery. Indeed, entire degree programs are being facilitated through the Web. Scholarship is ascertaining that the web-based environment is dramatically different from classroom-based instruction (Blythe, 2003; Boynton, 2002; Brady, 2001; Cargile-Cook & Grant-Davie, 2005; Depew, Fishman, Fahey Ruetnick, & Romberger, 2006; Hewett, 2004). The College Composition and Communication Conference Committee on Best Practices in Online Writing Instruction (2011) found that writing instructors use some different assignments online than they use in classroom-based settings (Jaschik, 2010). Likewise, instructional materials are taking a different shape.

Because the Web facilitates video and audio media, programs that certify web-based pedagogies such as Quality Matters encourage instructional materials that include video and slide shows. As mentioned in Chapter 6, there are not many studies focusing on neural processes related to cognition using slide shows. So this is an area that can be explored further from the neurobiology/physiology perspective to reinforce or challenge some conclusions reported in social science and rhetoric research. In particular, research may consider temporal synchronicity of visual and audio stimuli and how the position of visual information on a slide affects (or not) cognition. A debate in the field of visual rhetoric is how people "read" visual information. Studies have found that different people read web pages differently, so some study may focus on specific reading patterns relative to neurobiological dynamics. As I suggested in Chapter 6, research can consider particular attributes that affect cognition toward developing optimal slide shows.

Video can also be used extensively in web-based education. As acknowledged in Chapter 7, items such as camera angle and the amount of irrelevant information in a given screen can affect cognition. Studies in multimodal rhetoric tend to emphasize visual attributes of video, however, there is a growing corpus of scholarship that considers temporal synchronicity.

Consequently, research can help to develop effective instructional materials for web-based coursework. Studies into special education can inform this research as well. Students with cognitive neurological disorders such as autism need to have information provided to them in ways different than those who do not have such a disorder. As researchers study those with disorders (a biological-physical demographic characteristic), more can be ascertained about the multimodal combinations that work best to facilitate cognitive processes in such students.

Service-Learning/Internships

Service-learning pedagogy and internships have gained momentum as learning approaches because they put students in workplaces and in particular contextualized settings. Certain kinds of learning are context dependent, that is, certain information is better processed when provided in a specific context. Service learning/internships also happen to integrate multiple stimuli in the learning process. One listens, observes, and performs tasks, integrating multiple sensory experiences into the cognitive process of learning a given task.

However, certain tasks may be better learned in a sequence, much as I described relative to the air traffic control and pilot students. In each case, certain tools and instruction focused on parts of the responsibilities each kind of professional does to facilitate learning parts rather than the whole process at once.

Also, as suggested in the previous chapter in which I compared neurorhetoric associated with different media relative to a single task, the tools available to students will impact what media and modes can be used. A video product assumes that the user has access to an application that can display the video without experiencing difficulties. Slow Internet connections, for example, will render most videos useless because of modal interference that neurons will not be able to process well or filter. In such cases, the pdf that includes screenshots associated with certain steps may be better than a video.

WORKPLACE TRAINING

Workplace training comes in various modes and designs. While much involves hands-on training like that which one may experience in a service-learning or internship experience, it also integrates video and can include simulators like those described in Chapter 6. I suggest some approaches to researching workplace training in those kinds of environments here and how the model can inform such research.

Reward Neurons

Training involves learning. If employees understand that learning a given task may help them to earn a good performance review, a raise, or promotion, they are motivated to learn it. So, training materials should integrate explicit information to help the trainee understand the link between what he or she is about to learn, their job, and how learning it will help them attain a certain goal, be it a good performance review, which may be in a few weeks, or a promotion, which may occur next year.

Online Tutorials

Instructional videos like those I showed in Chapter 6 integrate several modes and sensory stimuli. There are several studies about rhetorical principles pertaining to video, and this model can help to enhance development of video products by encouraging developers to consider biological attributes that affect cognition. Studies of effectiveness of videos can use experimental designs as well as surveys to ascertain what participants perceive of their effectiveness.

Biological experiments using fMRI can measure blood flow as participants view videos and then perform particular tasks. Again, video facilitates mirror neuron activity to help one understand how to perform a task; fMRI technology can assess the degree to which mirror neurons are engaged with different modal combinations in video products.

Simulators

I detailed a few applications of simulators in Chapter 5. Simulators facilitate various degrees of immersion, which affects cognition and focus. The more immersed one is in an environment in which they will perform certain tasks, the more multisensory their learning experience can be, helping them to adjust to the actual environment.

As with video, the effectiveness of simulations can be assessed using experimental designs as well as focus groups. Further, fMRI studies can measure the degree to which immersion activates mirror neurons and the degree to which modal combinations in the simulations activate other neurons. Such information can help designers develop better instructional tools.

TECHNOLOGICAL LIMITATIONS OF RESEARCH TOOLS AND OPTIONS

As technologies develop, research can examine different physical activity as well. Much neurobiological research before 2000 used EEG technology to collect data, but the fMRI technology from which the hemo-neural hypothesis grew helps us understand that various systems in the brain are at work in facilitating cognitive processes. However, MRI technologies are still loud, which can affect any aural dynamics associated with multimodal analysis, perception, and even cognition. Until MRI equipment can reduce the noise level to minimally impact sensory attributes that contribute to perception and cognition, it is challenging to account for the noise in an analysis of the impact multimodal materials have on cognition.

While fMRI is able to identify a variety of patterns of neural activity, it may be difficult to account for and address in research the noise and any claustrophobic effects of using it for this research. A more viable approach may be magnetoencephalography (or MEG), which combines attributes of EEG and fMRI.

MRI machines are very loud and require immersion of the entire head and neck into them. Immersion would limit the ability for one to view any multimodal product, and the noise associated with the machine would practically render any audio associated with a multimodal product ineffective. MEG enables one to view multimodal products without being immersed completely. However, MEG also is not as fast as fMRI, and it measures different regions of the brain. So, data between the two tools may not agree with each other (Cohen & Halgren, 2004; Henson, Flandin, Friston, & Mattout, 2010).

As with any research tool, there are limitations associated with technologies available to measure neural activity. However, research must negotiate with these tools toward facilitating the best research possible.

PHARMACOLOGY

While this book has emphasized the neurorhetoric associated with multimodal instructional materials, research that I have identified can contribute to pharmacological developments such as developing drugs that treat neurological disorders that affect cognition. Medications that address disorders or conditions such as Attention-Deficit Hyperactivity Disorder (ADHD) focus on attributes of attention associated with cognition. Through the same research methods, studies that include analysis of brain activity and integrate the administration of medical drugs can help to ascertain which medications can most effectively treat cognitive problems and help improve cognition in various populations. Such studies may include understanding what structional materials may work in conjunction with certain medications or treatments.

Studies associated with the use of certain medications toward improving cognition or test performance tend to link the use of medications used for treating ADHD with better performance. ADHD has been linked to decreased dopamine levels (see, for example, Volkow et al., 2007). However, it is generally understood that ADHD medications such as Adderall and Ritilin work by increasing the number of neurotransmitters and prolonging their life in synapses. Neurotransmitters, like dopamine neurons, can be created out of amino acids, available from common diets. Categorized as amphetamines, they simply help one to pay attention by stimulating the cells that carry information across neural pathways. A problem with such medications, though, is that they are highly addictive and can lead to damage to neurons as well as other physical problems. If paying attention is the biggest hurdle to overcome, and the patient/subject would otherwise be able to understand a given concept or set of instructions, then such a medication can work well. However, cognition involves more than just paying attention.

Studies of those who have disorders such as autism can help physicians and teachers understand how people with such conditions process information. This understanding can lead to development of medications and learning materials that help such students and workers learn better and facilitate better cognitive function. Consequently, interdisciplinary study linking multimodal rhetoric scholarship and neuro-scientific scholarship can contribute to development of new medications and treatments. Biochemistry programs may team with technical communication or multimodal rhetoric programs to develop new curricula that encourage interdisciplinary study and work that contribute to such new treatments and medications.

As certain subjects with neural disorders are studied, the research can include administration of medications that affect neural activity, and fMRI and EEG studies can show how the medication is affecting processes toward cognition. I do not encourage a healthy person to be placed on medications just to improve their grades to compete better for certain academic or professional programs or careers. However, a combination of medication and learning with certain multimodal combinations may help those with certain neural disorders to learn better than they otherwise could.

Finally, studies are beginning to show a correlation between the activity in and around the heart and neural function: how general blood flow and neural activity around the heart affect cognition (McCraty et al., 2009; van der Wall & van Gilst, 2013). This suggests that studies could pursue how medications that affect the heart and blood flow, including cholesterol-related medications, generally affect cognition.

CHAPTER 10

Conclusion

As I indicated in Chapter 1, the cognitive experience includes biological as well as social dynamics. Cognitive science and the field of rhetoric generally recognize these attributes of cognition—social and biological attributes related to facilitating an understanding of our world. Disciplinary boundaries have compromised the discussion of these cognitive neuroscience dynamics. I have attempted to cross these boundaries in this book with this model. With a model that integrates social and neurobiological attributes of rhetoric to describe the ways our brain processes multimodal information toward cognition, I encourage further studies that combine these once-competing perspectives.

Rhetoric considers how an audience reacts relative to the way information is presented. One designs a message to facilitate a certain response from an audience. As Aristotle (1991) and Perelman and Olbrechts-Tyteca (1969) acknowledged, the way a message is conveyed must consider the audience's disposition in order to accomplish its purpose, and this disposition includes one's social disposition or biological/physical disposition. Indeed, Aristotle noted that this likely involves an audience that may have "limited intellectual scope and limited capacity to follow an extended chain of reasoning" (p. 76). If the audience's cognitive capacities are not considered in developing the message, the meaning of the message will be lost. This model attempts to take into consideration both biological and social dimensions of cognition.

Neural process studies help to explain some of the biological attributes connected to findings of multimodal scholarship, that is, they help us understand why, from a biological perspective, certain multimodal products facilitate a better understanding of information than other multimodal products.

The model I presented in Chapter 3 integrates what I consider to be the five attributes involved in cognition and that are affected by biological and social dynamics. The sixth attribute—medium—frames how one can design a message. This model is shown again in Figure 10.1.

As mentioned in previous chapters, the fields of rhetoric and neurobiology tend to examine multimodal and multisensory experiences differently, using different

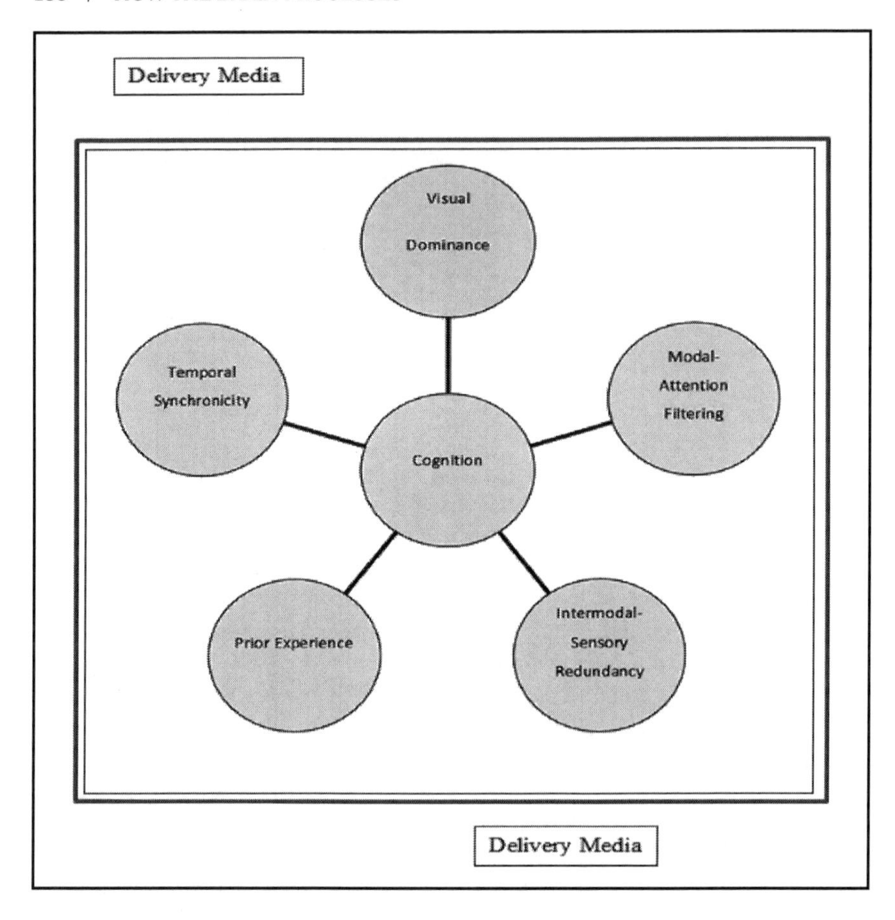

Figure 10.1. Model.

tools especially. While rhetoric studies tend to focus attention on composition and observed behaviors or surveys of audience perception of content, neurobiological studies focus on particular neuron behaviors inside the brain based on biomedical technologies. However, these neurobiological studies involve analysis of such activity relative to certain stimuli. Rhetoric can contribute to such studies, and the field of rhetoric can benefit from integrated studies likewise. Interdisciplinary research, facilitated by a model that engages the scholarship of multiple disciplines, can help with the development of better instructional designs and tools.

In each chapter of this book, in addition to citing specific studies, I alluded to varying degrees to events relatively common in our daily lives. Given how often we encounter multimodal information that our brain tries to process toward cognition, it is troubling to know that comprehensive study of such dynamics is hindered by the disciplinary divisions I have mentioned. The model I have presented in this book is a way to synthesize scholarship in the different fields productively.

Through synthesizing multimodal rhetorical theory with what is understood about how the mind processes information related to learning tasks and concepts relative to its use of blood, it is possible to refine that theory as it pertains specifically to training/instruction and process improvement, two topics of considerable interest in both industry and education. Examples of potential studies include ascertaining how the brain processes certain modal combinations; rhetoric scholars can design these combinations and use common approaches in social science research to contribute data to joint studies. Studies that include both biomedical technologies as well as social science research methods such as surveys and interviews with participants as well as observation and quasi-experimental designs can triangulate each other.

In Chapters 3 and 4, I detailed the new model, and I provided descriptive analyses of cases related to it in Chapters 5–8. Such analyses encourage further study via an interdisciplinary approach that can inform development of educational materials and workplace training. Chapter 5, in particular, featured a detailed case study of the Training Within Industry (TWI) program, which is esteemed presently in the lean operating and continuous improvement movements in business and industry. The other chapters considered digital media commonly used in education and training.

There are grants available through the National Institute for Humanities as well as foundations commonly associated with biomedical science research such as the National Science Foundation. Joint research can facilitate triangulation of information, leading to development of effective instructional materials that consider a population's disposition. Finally, I argue for further study into the neuroscience of learning and multimodal designs that facilitate learning and for a closer connection between neuroscience and multimodal rhetoric scholarship.

DIRECTIONS FOR FUTURE GROWTH

In the previous chapter, I identified ways that interdisciplinary research that integrated the model could enhance education and training as well as facilitate developments in pharmaceutical treatments. Here I suggest paths that scholarship may take to develop the model proposed in this book more fully. I mentioned in the Preface that the model is an early effort at finding a means by which interdisciplinary research can occur. I also indicated that the model is descriptive. It describes various attributes of cognition of any new information, but I recognize that it is not comprehensive; as I stated in the Preface, it is limited by my own disciplinary boundaries, some of which I have crossed with this work.

Because it is in its infancy, it has room to grow and develop. I encourage scholars to pursue development of it likewise, crossing their own disciplinary boundaries by conducting research with scholars in other disciplines or applying it to their own research, as I have, toward adding attributes to this model so it may represent a more comprehensive model of cognition.

Some questions emerge to guide further examination and development of the model. I ask researchers to consider their own disciplinary theories and how they may be integrated into the new model. I have characterized connections between the multimodal rhetoric and neurobiology of cognition, and I open the door to other consideration with these questions:

1. What principles of the model proposed here seem similar to attributes of theories of cognition in other field(s)?
2. Can such attributes be merged with the particular principles of the model to explain more fully that particular principle's relationship to cognition of new information?
3. Does a new nomenclature need to be developed, or can existing terms from different fields be used and synthesized, as the term "multimodal" was able to be synthesized between "multimodal rhetoric" and "multimodal integration," to characterize principles of multimodality here?
4. What attributes of cognition are not identified yet within the model? Does a new principle need to be added, or can one be developed further? That is, do some principles require subprinciples?

Particularly related to question 4, the principle of prior experience includes consideration of one's previous experiences with learning tools, previous knowledge of a concept, and previous learning approaches. However, the fields of language and linguistics invite further clarification of biological and social dynamics involved in cognition, especially relative to language disorders affected by neural traits and the social psychology of language. That is, can the principle of prior experience adequately address cultural attributes of language and cognition, or does a new principle need to be added to the model to address such attributes?

The kinds of interdisciplinary research that I have suggested can address these questions. Ultimately, the improvements made to instructional materials are what matters most, but an integrated model will enable researchers to consider the many attributes of cognition better toward developing those materials and improving learning.

APPENDIX A

Challenges in Historical Research of Multimodal Rhetoric and Cognition

Historical research in technical communication and cognition presents its own challenges. Specifically, the following six issues emerged as I worked on the historical study presented in this book:

1. Concerns about the memory of older adults who were interviewed;
2. Interview sampling relative to representation of the population;
3. Dealing with sensitive workplace information;
4. Coding multimodal documents;
5. Accounting for missing pages in archived documents; and
6. Ascertaining actual use of printed materials.

This study includes content analyses of interviews and archived documents. A majority of it relies on qualitative analyses, however, it also includes some quantitative analysis as well relative to readability dynamics. A recent special issue of *College Composition and Communication* (September 2012) dealt with issues in archival research. A few of the articles in that issue raised questions about what counts as archives and ethics of archival research. Gaillet (2012) raises several questions concerning archival research, including how to locate archives and ways to triangulate findings (p. 38). Many workplaces do not like for industry secrets to be reported in research studies, however, once a government record is declassified, it becomes open to the public for consideration. One way to locate new archives is to review entities that the government once owned or operated to ascertain any declassified materials. Also, my effort to use interviews and archival records is considered appropriate for triangulation.

Yin (1984) characterized case study research as an empirical inquiry that investigates a contemporary phenomenon within its real-life context, wherein the boundaries between phenomena and context are unclear and in which multiple sources of evidence are used (p. 23). MacNealy (1999) acknowledged that "case studies tend to rely heavily on interviews" (p. 203). Yin further acknowledged that, "one of the most important sources of case study information is the interview" (p. 88). Generally, interviews allow researchers to conduct open discussions about phenomena under study with participants.

Yin (1984) also stated that, "documentary evidence is likely to be relevant to every case study topic" (p. 85). In particular, he explained the usefulness of reviewing documents to triangulate information from other sources as well as to enhance that information (p. 86). Yin cautioned that researchers who use documents and archival records must try to understand the contexts in which the materials were developed and for what audiences (p. 88). So interviews can help ascertain those contexts.

Silverman (2006) pointed out that analysis of documents also facilitates an understanding of choices writers make regarding how to represent information for particular audiences (p. 152). Consequently, my coding includes study of patterns in the modes used to represent information in documents as well as reports of other modes of representation used to communicate information at the Arsenal. As a historical case study, I rely on analyses of documents that have been archived at the Arsenal as well as on data collected from interviews with members of the Fieldview community, some of whom worked at the Arsenal.

Interview Methods

Brandt (2001) explained that life-story research integrates "historical, sociological, psychological and phenomenological inquiry . . . [and includes] structured and less structured interviews" (p. 10). In this historical case study, I use a similar approach to Brandt's, interviewing those who lived the historical literate experience. Some of the interview participants worked at the Arsenal during the period of study, while other participants who did not work at the Arsenal lived in the community, attending the school district and participating in community organizations. These interviews reflect the participants' recollections of their practices. A concern about these interviews is the sampling/population associated with it.

RESEARCH ISSUE 1: HISTORICAL STUDIES—MEMORY AND ETHICAL RESEARCH

Many of the interview participants were over 80 years old at the time of the interview. Literature on concerns about memory related to interviews with elderly participants recommends tests of cognition and recall to ascertain problems with disorientation and dementia (e.g., Eeles & Rockwood, 2008; Taub, 1980). None of my participants had indications of disorientation or dementia, and people with whom I spoke about the participants acknowledged belief that the participants had a "sharp" memory.

Kirkevold and Bergland (2007) as well as Decker and Adamek (2004) acknowledged memory deficits as a concern in qualitative interviews with elderly participants. Both recommended interviewers take additional time to establish rapport with their participants and to use a less-structured interview, which Decker and Adamek acknowledged can yield "richer" and "more comprehensive" data than that gathered from structured interviews (p. 61). This could affect the accuracy of data. I attempted to address this concern by triangulating data sources: If a number of participants identified the same dynamics or experiences and practices, then I could conclude that the information was reliable.

RESEARCH ISSUE 2: INTERVIEW SAMPLING

The total population of possible participants is limited to people who lived in the particular community during that time frame and those who may or may not have worked at the Arsenal too. Consequently, my sample includes some who worked at the Arsenal and some who did not work there but lived in the community at some point during the period of study. However, because this is a historical study, and one must consider that people move to other places, and people may pass away, my sampling is limited to those who stayed in the area and were still living. This affects sampling relative to race and gender representation of the total population.

In attempting to facilitate a sampling that provided a cross-section of the population of the area during the time period, I used two approaches to recruit participants. I made announcements about my research project, inviting people to participate, at two of the local historical society meetings—one at a November meeting attended by approximately 100 people, and the second at the February meeting, attended by approximately 30 people. At each meeting, a speaker was scheduled to present information about local history. Prior to the presentation, society leaders permitted me to announce my project and facilitate volunteers. Generally, I introduced myself as a graduate student at a local institution who was studying the relationship the Arsenal had with the community, and I was focusing on reading and writing practices of people who worked at the Arsenal and/or lived in the community between 1940 and 1960. These announcements produced a list of 23 names.

I called anyone who wrote their name and contact information in response to these announcements or were referred to me. In each of these conversations, I acknowledged the nature of my study, what their participation would entail (an interview lasting approximately 30 minutes), and asked if they would be willing to participate. Most (18) of these people were willing to participate, and I made an appointment to interview them at their home, with IRB approval of the research protocol.

Race/Ethnicity

Because the sample was derived from announcements at a social gathering and referrals, this was a convenience sampling. Also, all of the volunteers from this initial set of announcements and calls were Caucasian. According to Rodabaugh (1975), African Americans composed approximately 2.5% of total war-related employment in 1940, and 8.2% of total employment in war-related industries in 1945. African Americans also worked at the Arsenal and lived in the community, however, while records documenting the total number of employees and their gender and ethnicity breakdowns were maintained, almost all of these records were lost or destroyed. The only record of this data that I could find was within the historical summary of July to December of 1943. According to this summary, African Americans (termed "negroes" in the archived documents) composed at most about 11% of the workforce there during that period.

I made an effort to include African Americans in the study, as I detail in the previously mentioned book, but to no avail.

Gender

The sampling also included more than twice as many females as males (13 females, 5 males). Women tend to outlive men by approximately 5 years: The average life expectancy is 80.5 years for White females and 75.3 years for White males (Shrestha, 2006). According to the 1943 historical summary of operations at the Arsenal, until September/October of 1942, males made up a large majority of the workforce at the Arsenal. However, in September and October of 1942, given the shrinking labor market, "experiments" were conducted to ascertain which production operations could be performed by women. "These tests demonstrated the fact that many operations heretofore thought too strenuous for female labor could be performed efficiently by them" (Shrestha, 2006, p. 313). So gender-related employment shifts, especially on load lines, where most of the assembling of munitions occurred.

The interview sampling included two females who worked at the Arsenal. Also, the interview questions invited participants to share information about practices that any of their relatives who worked at the arsenal shared with them in an effort to include more of the experiences of those who worked at the arsenal. One of the interviewees who did not work at the Arsenal spoke of an aunt who worked there. Consequently, the interview data included perspectives from female workers.

A total of 11 of the 18 interviewed were students in the Fieldview school district at different points during the period studied. Five worked at the Arsenal during the period studied and two worked in the community; one of these was a teacher in the Fieldview school district. Six (33.3%) of the participants were native to Fieldview prior to 1940, and five (27.8%) migrated to the area between 1940 and 1950. Six migrated to the area between 1950 and 1955, and one migrated to the area after 1955. Of the 12 who migrated, 5 migrated from within Ohio, while 4 migrated from more than one state away, and 9 of the 12 who migrated to the area did so for work-related reasons. Four migrated for work specifically at the Arsenal, and 11 participants reported that they, their parent(s), or a relative worked at the Arsenal. One participant was unable to recall what kind of work their relative did.

Generally, then, the interview sample included representation of people who were native to the area prior to the Arsenal's construction as well as those who migrated into the area for work at the Arsenal, those who worked at the Arsenal—male and female—and who held different positions there, and those who did not work at the arsenal but experienced community, home, and school-related literacy practices. However, a limitation of the study is that it did not include representation from African Americans.

Interview Questions

The interviews were semistructured to open-ended (Silverman, 2006). The script that I used for the interviews is in Appendix B. Most of the questions follow those Brandt used in her study (2001). While she asked her participants about various reading and writing practices, Brandt's questions were open-ended to allow participants to provide responses that included narratives of particular practices. I used similar questions because of the open-ended nature of the questions and because they allowed participants to speak freely of their experiences.

Generally, the script acted to guide the questioning, however, after ascertaining demographic information, the interviews tended to occur as conversations, with the participants responding at length to several questions. These responses often integrated information that could answer other questions in the set because of relationships between questions. For example, question 17 asked, "How much reading was required of the training program?" and question 19 asked, "How much reading was required in the job(s)? What kind?" Some participants included job-related reading requirements as they responded to question 17. Consequently, most of the interviews do`id not follow the script as it appeared. Sometimes a response facilitated a question not in the script as a follow-up to clarify or explain some information.

Unit of Analysis

Holsti (1969) identified six generally used units of analysis for content analysis: word or symbol, theme, character, sentence or paragraph or item (pp. 116–117). For this study, I used "theme" as my unit of analysis, which Holsti labeled as the most useful unit for content analysis (p. 116). Holsti characterized the "theme" unit as an assertion made about some subject. In using "theme," the researcher must break down a particular statement into a single theme or themes to facilitate tabulation into categories. I identify categories later, however, understanding theme as my primary unit for content analysis helps explain transcription and document selection approaches.

Categories and Coding

In this section, I explain coding of responses to these questions. Most questions were coded relative to the presence of certain practices relative to those themes identified above; for example, "home writing" is coded relative to the presence of "journals and diaries," "correspondence"; and "work reading" is coded relative to "correspondence," "reports," and "manuals." These were coded as nominal data (categorical data) rather than as interval data (indicating some quantitative value) to avoid suggesting any value of literacy level or skill associated with any particular practice. Nominal variables do not suggest any particular ordering, ranking, or value associated with them. If, for example, I were to format "home writing" as an interval or ratio variable and give "correspondence" a higher value than "diaries," it would suggest that skills associated with correspondence were valued more or are more rigorous than skills used for writing diaries. I list all categories and related codes in Appendix B.

In coding passages from transcripts, I considered the information a particular question sought and then looked for certain codes represented in the response. I did not count how many times a given code was acknowledged, only whether it was acknowledged. An example is from the exchange below:

> Me: Did you do any reading or writing outside of school? You mentioned the Methodist Church writing that you did. Anything at home or leisure reading newspapers?
>
> Steve: I did letters—did a lot of letter writing and those types of things . . . to grandparents and I did some reading but not a lot of reading. I was a worker—had a paper route and paper routes and did a lot of work through school. We didn't have a whole lot of money. We just did those types of things.

This exchange would result in my coding this participant's home writing activity as limited to "letter writing." I also coded for non-print-linguistic literate practices any participants identified, because I consider the New London Group's (1996) conception of multiliteracies and multiple modes of representation within the scope of the study. A few of the participants who worked at the Arsenal acknowledged the hands-on training associated with their work there, which I coded as visual, aural, and experiential (combining spatial and gestural modes identified by the New London Group).

An example of coding a relative's workplace literacy experience, from the same interview, is below:

Me: Okay, you mentioned that your father worked at the Arsenal and that he participated in the construction of it. What did he do at the Arsenal when he started there?

Steve: He was in supervision and supervision in various titles for number of years and he retired in 1982 after 42 years at Boomtown Arsenal.

For this exchange, I would code the work experience represented as the person's "parent," the position as "supervisor," and employment years as "over 40." If the participant shared information about the relative's experiences at the Arsenal, I included that information in the dataset, which would then provide a more comprehensive picture of literate practices at the Arsenal.

RESEARCH ISSUE 3: WORKPLACE STUDIES AND SENSITIVE INFORMATION

Because of the Arsenal's position in national defense and security policies associated with it, people who worked there may not have felt free to share literacy narratives about the workplace practices. There was a security policy that forbade workers from talking about their work there. So I needed to be able to ascertain sensitive information that interviewees may not have been willing to offer. I was able to address this through reviewing archived documents that had been declassified.

Not only can review of workplace documents triangulate interview data, it can also provide information that interview participants may not share. Interview data sheds light on recollections of general practices of each person; review of actual workplace documents helps to understand specific practices and expectations of writers and readers there and to triangulate interview data pertaining to the workplace practices.

Another issue connected to this point is that companies tend not to want industry secrets reported in research studies. However, the workplace I studied was a government-owned, contractor-operated site, which means that documents were available once declassified. Because they were produced during a time of war and concerned production innovations, they have been classified and hidden from researchers until released.

Document-Analysis Methods

I reviewed documents from the Arsenal to better understand what practices actually occurred. As I mentioned above relative to triangulating with interview data, reviewing as many different types of documents from the workplace provided another necessary data source to build a comprehensive picture of literacy practices there. I included in this corpus historical reports/summaries that operators of the Arsenal prepared at irregular intervals, depending on how active the site was; standard operating manuals; newsletters; incident reports; and building specifications. The total corpus included in this study numbers over 40 documents. In the next section, I detail data sources within the context of data collection. Review of the variety of documents sheds light on specific literacy practices across different levels of the organization.

Review of several documents at the worksite addressed the research question, What literacy practices occurred at the Arsenal? Krippendorff (2004) observed that researchers can determine an effective sampling size by reviewing a large corpus of the entire population and selecting a smaller corpus of texts that represent patterns across the larger corpus (p. 123). I explain my process later in this section.

The archivist at the Arsenal gave me access to an electronic database listing archived documents. However, she also acknowledged that this list was incomplete: Many documents had not yet been included in the system since she had been working on it for some 3 years at that point and had more to review. Documents not yet in the system were still disorganized, and part of her job was to bring some organization to the Arsenal's materials.

I studied 43 archival documents from the Arsenal to

1. Ascertain relationships between different modes used for representation and literacy expectations associated with them to address the research question regarding "What relationship between print-linguistic literacies and visual literacies existed at the workplace?" and
2. Ascertain literacy expectations through readability testing to address the research question, "What literacy practices were required in the workplace?"

I drew randomly from the listing of available documents based on the number of documents archived relative to different types of documents. That is, first I ascertained some parameters of the number of documents of a given type and differences in those numbers (e.g., manual/SOP versus routine reports), and I did not use any systematic means of drawing texts for the sample other than skimming the catalog and picking from those published during the period of study.

Sampling of Documents

Holsti (1969) acknowledged that sampling documents for content analyses generally occurs in a 2- or 3-stage process: identifying a list of sources, drawing a sampling of entire documents, and sampling a limited number of pages from within documents (p. 130). This study used all three of these stages for the analysis of archived documents.

Categories: Approaches

Development of categories to facilitate analysis is debated in the literature, especially regarding use of a grounded-theory approach, as indicated above (Crano & Brewer, 2002; Holsti, 1969; Krippendorf, 2004). Generally, one can use an *a priori* approach identifying categories prior to the actual collection of data; or one can use a grounded-theory approach, letting categories that facilitate analysis emerge as one collects data. Strauss (1987) acknowledged that such analytical categories can be identified by the researcher at any of several phases in the research. Krippendorff (2004) acknowledged five different ways to define the unit(s) of study for content analysis: physical distinctions, syntactical distinctions, categorical distinctions, propositional distinctions, and thematic distinctions (pp. 103–109). Two that work for the study of texts that include print-linguistic and visual modes of representation, multimodal representation, are physical distinctions and categorical distinctions. Because this study examined such relationships, I used similar categories. Categories related to physical attributes of a message that the literature typically identifies, though, include amount of space on a page devoted to the object(s) of study, size of the object being studied on a given page, and frequency of occurrences of a given attribute (Crano & Brewer, 2002; Krippendorff, 2004; Matthiessen, 2007; Royce, 2007). Krippendorff (2004) also acknowledged that frequency of occurrences can characterize categorical distinctions—the frequency that a given category of a variable occurs suggests something about its value. While different documents may use different relationships relative to their purpose, examination of a single kind of document or modes of representation across different readerships relative to a single purpose may find particular trends.

While little debate about *a priori* categories exists, Holsti (1969) and Krippendorff (2004) encouraged a grounded-theory approach, while Crano and Brewer (2002) discouraged it. Crano and Brewer acknowledged that content analysis is a form of observational research, which necessitates identification of units prior to making observations (p. 247). Consequently, as acknowledged previously, I used mostly *a priori* categories. However, Holsti (1969) encouraged a grounded-theory approach, acknowledging that the categories must reflect the research questions and that the standardization of a set of categories assumes a large corpus of research on a given phenomenon (pp. 101–102). Further, Krippendorff (2004) suggested that a grounded-theory approach is appropriate when he stated that units "emerge in processes of reading and thus implicate the experiences of the analyst as a competent reader" (p. 98). As noted above, Strauss (1987) explained that attributes of a grounded-theory approach can occur at each step in the research process (pp. 25–32). I used it principally within the concept-indicator phase, which Strauss acknowledged directs coding of certain empirical indicators (p. 25). Indicators are data that are indicators of a given "concept the analyst derives from them" (p. 25). Strauss went on to explain that through comparing similarities and differences across indicators/data, a category emerges, and related codes can be refined accordingly (p. 25). While most of the categories I used have been identified to some degree in literature previously acknowledged, I observed two important attributes of the documents and used a grounded-theory approach toward accounting for those items, consequently, my coding reflects certain issues within historical studies not presented in the literature. Specifically, placement of certain graphics and the presence of any missing pages became important to document. These are two more issues I needed to address.

RESEARCH ISSUE 4: CATEGORIES FOR WORKPLACE MULTIMODAL PRINT MATERIALS

The categories I used differ somewhat from those generally identified in the literature, because the literature on multimodal pages tends to deal with content analyses of newspapers/magazines, which tend to use column space as a measure of use. Fewer studies have analyzed workplace reports or manuals. In addition to several commonly used categories, two additional categories that emerged pertinent to aggregate data are that of location of visuals relative to the text that describes them: "co-location" (or relative location) and "missing items."

The more recent work in multimodal analysis seems to examine relationships between text and moving images/animation on a computer screen (e.g., Wysocki, 2001) or comparing text-based representations to animated representations of the same content (e.g., Unsworth, 2007). Further, Unsworth (2007) acknowledged that certain kinds of visual representations may articulate explicit semiotic relations between the visual and the viewer or the writer and reader (p. 332). For example, a photograph showing actual people who represent the viewer will be more meaningful to that viewer than a diagram showing some representation of the viewer (p. 348). So, I coded for photographs as well as drawings or diagrams, because a photograph may carry more rhetorical meaning for a reader in a certain context—a manual, for example. Use of a lot of photographs compared to any other kind of graphic may suggest a rhetorical decision by the writer associated with the reader's literate background.

I also coded these graphic and textual distinctions because Matthiessen (2007) acknowledged that combinations of graphics and text have been "a feature of literate cultures," but that the division of labor between images and text has undergone "significant" changes over time (p. 29). An emphasis on diagrams and images and photographs suggests an emphasis on visual literacy skills as a theme, while an emphasis on print-linguistic text suggests an emphasis on print-linguistic literacies as a theme.

Bazerman (2008) asserted that "site-specific questions must attend to the particular character, opportunities, and difficulties of gathering data at the site as well as to the kind of analysis the data will allow" (p. 306). Consequently, my data-collection methods attempt to address specific site-related issues, like missing pages in documents, lack of color photography, and age/memory-related issues associated with interview participants. While Brandt (2001) did not examine any textual materials, Heath (1983), Street (1984) and Scribner and Cole (1981) considered texts within their design to understand relationships between literate practices in different settings. Generally, for example, Heath coded such documents relative to the kind of verbal and visual representations associated with them, though she did so very generally, and she used narrative analysis. For example, she observed that, "reading material in the mill, beyond section names and signs marking restrooms, lunchroom, trash cans, soft drink machines, etc., is limited to information on the bulletin boards" (p. 234). She did not report an empirical textual analysis showing frequencies of certain relationships or types of text or graphics. However, I attempted to code these variables for empirical study here to facilitate mixed-methods analyses because of potential relationships they may reveal between literacy expectations of readers and the material presented in the documents. This analysis is consistent with the consideration of multiple modes of representation that is associated with this study. Extensive use of visual representations of information in documents suggests an emphasis on visual literacy skills, while extensive use of print-linguistic text in documents suggests an emphasis on print-linguistic literacy skills.

It became clear that visuals in manuals and SOPs produced after 1950 were positioned after the text information, while they were placed within the related textual information frequently in such documents produced before 1950. I included the relative location code because of the various placements that I observed. Bateman, Delin, and Henschel (2007) considered such positioning of information within their category of "rhetorical structure": "how the content is divided into . . . main material, supporting material" (p. 155). Generally, appendixed information is considered supplemental material, while graphics placed within the text are considered primary to the purpose of the document (Markel, 2010; Oliu, Brusaw, & Alred, 2010). Because of this temporal difference in locating graphics and the potential association with perceptions of the relationship between the text and graphics, I coded for these different locations.

RESEARCH ISSUE 5: ACCOUNTING FOR MISSING PAGES

Some pages were missing from some documents, having been destroyed or lost; and some of these pages contained graphics, according to references in pages that exist. Some of these were textual pages, while others were appendixed pages. For example, some documents included textual references to graphics found in appendices, and the appendices were missing.

Rather than omit these documents entirely from analysis, I attempted to include the fact that some pages were missing in the analysis while including the materials that did exist. Excluding the entire document from the analysis and including only documents that were complete omitted a portion of the population of documents. I tried to identify frequency of certain phenomena, such as the number of text-only and graphics-only and combination pages as well as types of graphics used. A given document that was missing pages may have had more of certain types of pages than I could observe, but I could analyze only the material I observed. So coding for missing pages lets the reader understand that there may have been more graphics or text-only pages or more of a certain type of graphic present, but that I could not document it. In any case, I tried to minimize the number of missing pages included in the data so as to minimize the effect those missing pages had on the analysis.

Readability Tests

Finally, readability tests are part of content analyses (Crano & Brewer, 2002, p. 262; Holsti, 1969, p. 89; Krippendorff, 2008, p. 58) because they are a measure of reading skill expected of a given audience. Because one of the related research questions in this study is to understand literacy requirements of employees who worked at the Arsenal relative to how their background may have affected literate practices, I included analysis of literacy levels expected of and practiced by employees as evidenced in readability tests of various documents. To facilitate this analysis, I sampled three to four passages from a selection of each type of document—manuals/SOP, routine report, special report, newsletter—randomly. That is, I turned groups of pages in no systematic fashion, coming to a page that may have had graphics on it or not, and selecting a passage from that page. I drew this sampling in this way to establish a set that represented readability attributes for the entire document.

I retyped two or three entire articles (newsletter) or passages (manuals, reports, building specifications) of an average of over 200 words per passage, with a range of 70 words to 422 words. I then applied five different readability tests, all of which measured the grade level necessary to understand the content using different algorithms to each sampled passage. To run the test, I used the website http://www.online-utility.org/english/readability_test_and_improve.jsp. This website allows the user to copy and paste text passages into a box, and it will calculate results for several different readability tests.

The site facilitates several tests, including Gunning-Fog, Coleman-Liau, Flesch-Kincaid, ARI, and SMOG. All of these are associated with measures of readability relative to a particular grade level. That is, the resulting output number associated with any of these tests reflects the grade level needed for a reader to be able to understand the passage. However, each uses a different set of variables to arrive at the grade-output.

Both Krippendorff (2004) and Crano and Brewer (2002) acknowledged that the Flesch-Kincaid test is the most-used test, acknowledging that even the U.S. Department of Defense uses it to measure readability of its documents; consequently, I reported only data related to the Flesch-Kincaid test in this study. This test, which is derived from a test Flesch developed in 1943, considers the average number of words used per sentence and the average number of syllables per word. Examination of the mean grade-level readability score associated with the various documents in the workplace helps to identify patterns of reading expectations of targeted audiences of those documents. If the mean of a particular kind of document is lower than for another document, this suggests that the audience of the first kind of document is expected to have a lower grade-level reading skill than readers of the second document.

RESEARCH ISSUE 6: ASCERTAINING ACTUAL USE OF DOCUMENTS

While I conducted content analyses of several documents, a problem that occurred to me is that just because documents exist in a workplace, employees may not actually read them, or they may use them a certain way not stated within the documents themselves. A researcher may find information regarding a document's intended audience and purpose within the document, or they may be able to infer by nature of information in it. However, how it was read is difficult to ascertain with just the documents.

To address this problem, I asked people whom I interviewed and who worked at the site to what degree and how people used the actual documents (source triangulation). Included in my interview sampling were 11 people who worked at the site or whose relatives worked there. According to Krippendorff (2004), interviews allow participants to recall their interaction with various documents and identify which were of primary importance for them and how they interacted with any workplace documents associated with various tasks.

Qualitative Analyses

I used qualitative analytical methods in this study. Yin (1984) encouraged analysts to identify a particular theme or unit on which to facilitate analysis of case study data. He generally discouraged use of statistical analyses within case study design—single case study and multiple case study—largely because variables studied are unlikely to have any "variance" (p. 113) and selection of particular cases is "not based on any sampling logic" (p. 124). However, he encouraged using a "pattern-matching" approach to analysis to facilitate internal validity in a single case study and replication of a given study to other cases, thereby enhancing external validity. Data from different observations within a case study may show a certain pattern from which findings and conclusions are drawn.

Traditionally, analyses of content tend to focus on the presence of given phenomena being studied and patterns that emerge. Within this study, I analyzed patterns and trends of certain phenomena. Further, literature that describes analyses of visual information tends to call attention to certain categories of attributes of graphic data relative to salience. These include use of color, relative size, richness, and sharpness (Kress & van Leeuwen, 2006; Rowley-Joliet, 2004; Tufte, 2006; van Leeuwen, 2003). Consequently, my coding included relative size and the type of graphic (e.g., photograph versus diagram). Color photography and printing were not available for part of the period under study, so I did not include them as a characteristic of study. Bateman et al. (2007) considered this dynamic in their category "production constraints," which may be attributed to availability of color photography and/or macro-economic concerns about costs of using such (p. 155). Further, because I examined documents published over a period of 20 years, I also observed changes that occurred in their content and format over this period. Several documents published in a given 2- or 3-year time frame exhibited similar attributes; and the attributes differed with documents published more than 5 years apart. I discussed these progressive changes over time as shifts in attributes and general trends.

CONCLUSION

The sample for the document analyses is a good representation of the range of literacy practices at the Arsenal during the period of the study. Counting frequency of occurrence, testing readability, and measuring means of the categories involved allowed for an understanding of literacy practices and expectations at the Arsenal. Also, interviews with several members of the community, including former employees of the arsenal, helped to triangulate data associated with the interviews as well as data from archived documents.

APPENDIX B

Interview Questions

1. Were you born in the Windham area or at what age did you move to the area?
2. School(s) attended?
3. College? Degrees attained: what, when completed?
4. Age when you began working in area?

General Adult education

5. Did you attend any adult education/training classes/workshops?
6. When?
7. Where?

Adult education—transition-community-sponsored training

8. What training programs do you recall were available to help with your or your family's transition to life in northeast Ohio?
9. What community-based training programs were available to develop reading and/or writing skills?
10. What reading skills were developed in this program?
11. What writing skills were developed in this program?
12. How did the skills developed in this program help you or your relative(s) at home, education and/or in any work you or they have done?

Job search

13. Did you or anyone in your family work at the Arsenal? What position(s)?
14. Did anyone help you complete job applications?
15. What questions related to reading/writing skills were asked in any job interviews?

Work-related skills

16. What was the Arsenal's training program like?
17. How much reading was required of the training program?
18. How much writing was required of the training program?
19. How much reading was required in the job(s)? What kind?
20. How much writing was required in the job(s)? What kind?

Home/other work literacy

21. How did the skills you/your relatives learned in this program help at home or in any other job you/they've held?
22. What other reading and writing training did you/they receive?
23. How did this training help you/them at home or at any other job you/they've held?

APPENDIX C

The Arsenal of Democracy Speech: December 29, 1940

Franklin Delano Roosevelt

Retrieved from
http://www.americanrhetoric.com/speeches/fdrarsenalofdemocracy.html
American Rhetoric: Top 100 Speeches.
March 14, 2008

My friends:

This is not a fireside chat on war. It is a talk on national security; because the nub of the whole purpose of your President is to keep you now, and your children later, and your grandchildren much later, out of a last-ditch war for the preservation of American independence, and all of the things that American independence means to you and to me and to ours.

Tonight, in the presence of a world crisis, my mind goes back eight years to a night in the midst of a domestic crisis. It was a time when the wheels of American industry were grinding to a full stop, when the whole banking system of our country had ceased to function. I well remember that while I sat in my study in the White House, preparing to talk with the people of the United States, I had before my eyes the picture of all those Americans with whom I was talking. I saw the workmen in the mills, the mines, the factories, the girl behind the counter, the small shopkeeper, the farmer doing his spring plowing, the widows and the old men wondering about their life's savings. I tried to convey to the great mass of American people what the banking crisis meant to them in their daily lives.

Tonight, I want to do the same thing, with the same people, in this new crisis which faces America. We met the issue of 1933 with courage and realism. We face this new crisis, this new threat to the security of our nation, with the same courage and realism. Never before since Jamestown and Plymouth Rock has our American civilization been in such danger as now. For on September 27th, 1940—this year—by an agreement signed in Berlin, three powerful nations, two in Europe and one in Asia, joined themselves together in the threat that if the United States of America interfered with or blocked the expansion program of these three nations—a program aimed at world control—they would unite in ultimate action against the United States.

The Nazi masters of Germany have made it clear that they intend not only to dominate all life and thought in their own country, but also to enslave the whole of Europe, and then to use the resources of Europe to dominate the rest of the world. It was only three weeks ago that their leader stated this: "There are two worlds that stand opposed to each other." And then in defiant reply to his opponents he said this: "Others are correct when they say: With this world we cannot ever reconcile ourselves. I can beat any other power in the world." So said the leader of the Nazis.

In other words, the Axis not merely admits but the Axis proclaims that there can be no ultimate peace between their philosophy—their philosophy of government—and our philosophy of government. In view of the nature of this undeniable threat, it can be asserted, properly and categorically, that the United States has no right or reason to encourage talk of peace until the day shall come when there is a clear intention on the part of the aggressor nations to abandon all thought of dominating or conquering the world.

At this moment the forces of the States that are leagued against all peoples who live in freedom are being held away from our shores. The Germans and the Italians are being blocked on the other side of the Atlantic by the British and by the Greeks, and by thousands of soldiers and sailors who were able to escape from subjugated countries. In Asia the Japanese are being engaged by the Chinese nation in another great defense. In the Pacific Ocean is our fleet.

Some of our people like to believe that wars in Europe and in Asia are of no concern to us. But it is a matter of most vital concern to us that European and Asiatic war-makers should not gain control of the oceans which lead to this hemisphere. One hundred and seventeen years ago the Monroe Doctrine was conceived by our government as a measure of defense in the face of a threat against this hemisphere by an alliance in Continental Europe. Thereafter, we stood guard in the Atlantic, with the British as neighbors. There was no treaty. There was no "unwritten agreement." And yet there was the feeling, proven correct by history, that we as neighbors could settle any disputes in peaceful fashion. And the fact is that during the whole of this time the Western Hemisphere has remained free from aggression from Europe or from Asia.

Does anyone seriously believe that we need to fear attack anywhere in the Americas while a free Britain remains our most powerful naval neighbor in the Atlantic? And does anyone seriously believe, on the other hand, that we could rest easy if the Axis powers were our neighbors there? If Great Britain goes down, the Axis powers will control the Continents of Europe, Asia, Africa, Austral-Asia, and the high seas. And they will be in a position to bring enormous military and naval resources against this hemisphere. It is no exaggeration to say that all of us in all the Americas would be living at the point of a gun—a gun loaded with explosive bullets, economic as well as military. We should enter upon a new and terrible era in which the whole world, our hemisphere included, would be run by threats of brute force. And to survive in such a world, we would have to convert ourselves permanently into a militaristic power on the basis of war economy.

Some of us like to believe that even if Britain falls, we are still safe, because of the broad expanse of the Atlantic and of the Pacific. But the width of those oceans is not what it was in the days of clipper ships. At one point between Africa and Brazil the distance is less than it is from Washington to Denver, Colorado, five hours for the latest type of bomber. And at the north end of the Pacific Ocean, America and Asia almost touch each other. Why, even today we have planes that could fly from the British Isles to New England and back again without refueling. And remember that the range of the modern bomber is ever being increased.

During the past week many people in all parts of the nation have told me what they wanted me to say tonight. Almost all of them expressed a courageous desire to hear the plain truth about the gravity of the situation. One telegram, however, expressed the attitude of the small minority who want to see no evil and hear no evil, even though they know in their hearts that evil exists. That telegram begged me not to tell again of the ease with which our American cities could be bombed by any hostile power which had gained bases in this Western Hemisphere. The gist of that telegram was: "Please, Mr. President, don't frighten us by telling us the facts." Frankly and definitely there is danger ahead—danger against which we must prepare. But we well know that we cannot escape danger, or the fear of danger, by crawling into bed and pulling the covers over our heads.

Some nations of Europe were bound by solemn nonintervention pacts with Germany. Other nations were assured by Germany that they need never fear invasion. Nonintervention pact or not, the fact remains that they were attacked, overrun, thrown into modern slavery at an hour's notice—or even without any notice at all. As an exiled leader of one of these nations said to me the other day, "The notice was a minus quantity. It was given to my government two hours after German troops had poured into my country in a hundred places." The fate of these nations tells us what it means to live at the point of a Nazi gun.

The Nazis have justified such actions by various pious frauds. One of these frauds is the claim that they are occupying a nation for the purpose of "restoring order." Another is that they are occupying or controlling a nation on the excuse that they are "protecting it" against the aggression of somebody else. For example, Germany has said that she was occupying Belgium to save the Belgians from the British. Would she then hesitate to say to any South American country: "We are occupying you to protect you from aggression by the United States?" Belgium today is being used as an invasion base against Britain, now fighting for its life. And any South American country, in Nazi hands, would always constitute a jumping off place for German attack on any one of the other republics of this hemisphere.

Analyze for yourselves the future of two other places even nearer to Germany if the Nazis won. Could Ireland hold out? Would Irish freedom be permitted as an amazing pet exception in an unfree world? Or the islands of the Azores, which still fly the flag of Portugal after five centuries? You and I think of Hawaii as an outpost of defense in the Pacific. And yet the Azores are closer to our shores in the Atlantic than Hawaii is on the other side.

There are those who say that the Axis powers would never have any desire to attack the Western Hemisphere. That is the same dangerous form of wishful thinking which has destroyed the powers of resistance of so many conquered peoples. The plain facts are that the Nazis have proclaimed, time and again, that all other races are their inferiors and therefore subject to their orders. And most important of all, the vast resources and wealth of this American hemisphere constitute the most tempting loot in all of the round world.

Let us no longer blind ourselves to the undeniable fact that the evil forces which have crushed and undermined and corrupted so many others are already within our own gates. Your government knows much about them and every day is ferreting them out. Their secret emissaries are active in our own and in neighboring countries. They seek to stir up suspicion and dissension, to cause internal strife. They try to turn capital against labor, and vice versa. They try to reawaken long slumbering racial and religious enmities which should have no place in this country. They are active in every group that promotes intolerance. They exploit for their own ends our own natural abhorrence of war. These trouble-breeders have but one purpose. It is to divide our people, to divide them into hostile groups and to destroy our unity and shatter our will to defend ourselves.

There are also American citizens, many of them in high places, who, unwittingly in most cases, are aiding and abetting the work of these agents. I do not charge these American citizens with being foreign agents. But I do charge them with doing exactly the kind of work that the dictators want done in the United States. These people not only believe that we can save our own skins by shutting our eyes to the fate of other nations. Some of them go much further than that. They say that we can and should become the friends and even the partners of the Axis powers. Some of them even suggest that we should imitate the methods of the dictatorships. But Americans never can and never will do that.

The experience of the past two years has proven beyond doubt that no nation can appease the Nazis. No man can tame a tiger into a kitten by stroking it. There can be no appeasement with ruthlessness. There can be no reasoning with an incendiary bomb. We know now that a nation can have peace with the Nazis only at the price of total surrender. Even the people of Italy have been forced to become accomplices of the Nazis; but at this moment they do not know how soon they will be embraced to death by their allies.

The American appeasers ignore the warning to be found in the fate of Austria, Czechoslovakia, Poland, Norway, Belgium, the Netherlands, Denmark, and France. They tell you that the Axis powers are going to win anyway; that all of this bloodshed in the world could be saved, that the United States might just as well throw its influence into the scale of a dictated peace and get the best out of it that we can. They call it a "negotiated peace." Nonsense! Is it a negotiated peace if a gang of outlaws surrounds your community and on threat of extermination makes you pay tribute to save your own skins? For such a dictated peace would be no peace at all. It would be only another armistice, leading to the most gigantic armament race and the most devastating trade wars in all history. And in these contests the Americas would offer the only real resistance to the Axis power. With all their vaunted efficiency, with all their parade of pious purpose in this war, there are still in their background the concentration camp and the servants of God in chains.

The history of recent years proves that the shootings and the chains and the concentration camps are not simply the transient tools but the very altars of modern dictatorships. They may talk of a "new order" in the world, but what they have in mind is only a revival of the oldest and the worst tyranny. In that there is no liberty, no religion, no hope. The proposed "new order" is the very opposite of a United States of Europe or a United States of Asia. It is not a government based upon the consent of the governed. It is not a union of ordinary, self-respecting men and women to protect themselves and their freedom and their dignity from oppression. It is an unholy alliance of power and pelf to dominate and to enslave the human race.

The British people and their allies today are conducting an active war against this unholy alliance. Our own future security is greatly dependent on the outcome of that fight. Our ability to "keep out of war" is going to be affected by that outcome. Thinking in terms of today and tomorrow, I make the direct statement to the American people that there is far less chance of the United States getting into war if we do all we can now to support the nations defending themselves against attack by the Axis than if we acquiesce in their defeat, submit tamely to an Axis victory, and wait our turn to be the object of attack in another war later on.

If we are to be completely honest with ourselves, we must admit that there is risk in any course we may take. But I deeply believe that the great majority of our people agree that the course that I advocate involves the least risk now and the greatest hope for world peace in the future.

The people of Europe who are defending themselves do not ask us to do their fighting. They ask us for the implements of war, the planes, the tanks, the guns, the freighters which will enable them to fight for their liberty and for our security. Emphatically, we must get these weapons to them, get them to them in sufficient volume and quickly enough so that we and our children will be saved the agony and suffering of war which others have had to endure.

Let not the defeatists tell us that it is too late. It will never be earlier. Tomorrow will be later than today.

Certain facts are self-evident.

In a military sense Great Britain and the British Empire are today the spearhead of resistance to world conquest. And they are putting up a fight which will live forever in the story of human gallantry. There is no demand for sending an American expeditionary force outside our own borders. There is no intention by any member of your government to send such a force. You can therefore, nail, nail any talk about sending armies to Europe as deliberate untruth. Our national policy is not directed toward war. Its sole purpose is to keep war away from our country and away from our people.

Democracy's fight against world conquest is being greatly aided, and must be more greatly aided, by the rearmament of the United States and by sending every ounce and every ton of munitions and supplies that we can possibly spare to help the defenders who are in the front lines. And it is no more un-neutral for us to do that than it is for Sweden, Russia, and other nations near Germany to send steel and ore and oil and other war materials into Germany every day in the week.

We are planning our own defense with the utmost urgency, and in its vast scale we must integrate the war needs of Britain and the other free nations which are resisting aggression. This is not a matter of sentiment or of controversial personal opinion. It is a matter of realistic, practical military policy, based on the advice of our military experts who are in close touch with existing warfare. These military and naval experts and the members of the Congress and the Administration have a single-minded purpose: the defense of the United States.

This nation is making a great effort to produce everything that is necessary in this emergency, and with all possible speed. And this great effort requires great sacrifice. I would ask no one to defend a democracy which in turn would not defend every one in the nation against want and privation. The strength of this nation shall not be diluted by the failure of the government to protect the economic well-being of its citizens. If our capacity to produce is limited by machines, it must ever be remembered that these machines are operated by the skill and the stamina of the workers.

As the government is determined to protect the rights of the workers, so the nation has a right to expect that the men who man the machines will discharge their full responsibilities to the urgent needs of defense. The worker possesses the same human dignity and is entitled to the same security of position as the engineer or the manager or the owner. For the workers provide the human power that turns out the destroyers, and the planes, and the tanks. The nation expects our defense industries to continue operation without interruption by strikes or lockouts. It expects and insists that management and workers will reconcile their differences by voluntary or legal means, to continue to produce the supplies that are so sorely needed. And on the economic side of our great defense program, we are, as you know, bending every effort to maintain stability of prices and with that the stability of the cost of living.

Nine days ago I announced the setting up of a more effective organization to direct our gigantic efforts to increase the production of munitions. The appropriation of vast sums of money and a well-coordinated executive direction of our defense efforts are not in themselves enough. Guns, planes, ships and many other things have to be built in the factories and the arsenals of America. They have to be produced by workers and managers and engineers with the aid of machines which in turn have to be built by hundreds of thousands of workers throughout the land. In this great work there has been splendid cooperation between the government and industry and labor. And I am very thankful.

American industrial genius, unmatched throughout all the world in the solution of production problems, has been called upon to bring its resources and its talents into action. Manufacturers of watches, of farm implements, of Linotypes and cash registers and automobiles, and sewing machines and lawn mowers and locomotives, are now making fuses and bomb packing crates and telescope mounts and shells and pistols and tanks.

But all of our present efforts are not enough. We must have more ships, more guns, more planes—more of everything. And this can be accomplished only if we discard the notion of "business as usual." This job cannot be done merely by superimposing on the existing productive facilities the added requirements of the nation for defense. Our defense efforts must not be blocked by those who fear the future consequences of surplus plant capacity. The possible consequences of failure of our defense efforts now are much more to be feared. And after the present needs of our defense are past, a proper handling of the country's peacetime needs will require all of the new productive capacity, if not still more. No pessimistic policy about the future of America shall delay the immediate expansion of those industries essential to defense. We need them.

I want to make it clear that it is the purpose of the nation to build now with all possible speed every machine, every arsenal, every factory that we need to manufacture our defense material. We have the men, the skill, the wealth, and above all, the will. I am confident that if and when production of consumer or luxury goods in certain industries requires the use of machines and raw materials that are essential for defense purposes, then such production must yield, and will gladly yield, to our primary and compelling purpose.

So I appeal to the owners of plants, to the managers, to the workers, to our own government employees to put every ounce of effort into producing these munitions swiftly and without stint. With this appeal I give you the pledge that all of us who are officers of your government will devote ourselves to the same whole-hearted extent to the great task that lies ahead.

As planes and ships and guns and shells are produced, your government, with its defense experts, can then determine how best to use them to defend this hemisphere. The decision as to how much shall be sent abroad and how much shall remain at home must be made on the basis of our overall military necessities.

We must be the great arsenal of democracy.

For us this is an emergency as serious as war itself. We must apply ourselves to our task with the same resolution, the same sense of urgency, the same spirit of patriotism and sacrifice as we would show were we at war.

We have furnished the British great material support and we will furnish far more in the future. There will be no "bottlenecks" in our determination to aid Great Britain. No dictator, no combination of dictators, will weaken that determination by threats of how they will construe that determination. The British have received invaluable military support from the heroic Greek Army and from the forces of all the governments in exile. Their strength is growing. It is the strength of men and women who value their freedom more highly than they value their lives.

I believe that the Axis powers are not going to win this war. I base that belief on the latest and best of information.

We have no excuse for defeatism. We have every good reason for hope—hope for peace, yes, and hope for the defense of our civilization and for the building of a better civilization in the future. I have the profound conviction that the American people are now determined to put forth a mightier effort than they have ever yet made to increase our production of all the implements of defense, to meet the threat to our democratic faith.

As President of the United States, I call for that national effort. I call for it in the name of this nation which we love and honor and which we are privileged and proud to serve. I call upon our people with absolute confidence that our common cause will greatly succeed.

References

Allman, B., Keniston, L., & Meredith, M. A. (2009). Not just for multimodal neurons anymore: The contribution of unimodal neurons to cortical multisensory processing. *Brain Topography, 21*, 157–167.

Allman, B., & Meredith, A. M. (2007). Multisensory processing in "unimodal" neurons: Cross-modal subthreshold auditory effects in cat extrastriate visual cortex. *Journal of Neurophysiology, 98,* 545–549.

Amerine, R., & Bilmes, J. (1990). Following instructions. In M. Lynch & S. Woolgar (Eds.), *Representation in scientific practice* (pp. 323–336). Cambridge, MA: MIT Press.

Ammunition General, TM 9-1900 War Department Technical Manual. 1945.

Aristotle. (1991). *The art of rhetoric* (H. C. Lawson-Tancred, Trans.). London, UK: Penguin.

Arnheim, R. (1969). *Visual thinking*. Berkeley, CA: University of California Press.

Atlas Powder Company. (1943). *Semiannual summary of operations: Boomtown Army Ammunition Plant (AAP).* AAP Repository Documents.

Atlas Powder Company. (1944). *History of the operating contractor's organization and operation of the Boomtown Ordinance Plant* (Vol. 1). Wilmington, DE: U.S. Government.

Atlas Powder Company. (1944). *Semiannual summary of operations: Army Ammunition Plant (AAP).* AAP Repository Documents.

Atlas Powder Company. (1945). *Semiannual summary of operations: Ravenna Army Ammunition Plant (AAP).* AAP Repository Documents.

Azar, B. (2010). More powerful persuasion. *Monitor on Psychology, 41*, pp. 36–38. Retrieved June 3, 2013 from http://www.scn.ucla.edu/pdf/Persuasion-Monitor-2010.pdf

Baddeley, A. D. (1986). *Working memory*. Oxford, UK: Oxford University Press.

Ball, C. (2006). Designerly ≠ Readerly: Re-assessing multimodal and new media rubrics for use in writing studies. *Convergence: The International Journal of Research into New Media Technologies, 12,* 393–412.

Bateman, J., Delin, J., & Henschel, R. (2007). Mapping the multimodal genres of traditional and electronic newspapers. In T. D. Royce & W. L. Bowcher (Eds.), *New directions in the analysis of multimodal discourse* (pp. 147–172). Mahwah, NJ: Lawrence Erlbaum Associates.

Bazerman, C. (2008). Theories of the middle range in historical studies of writing practice. *Written Communication, 25,* 298–318.

Berlucchi, G., & Buchtel, H. A. (2009). Neuronal plasticity: Historical roots and evolution of meaning. *Experimental Brain Research, 192*(3), 307–319. doi: 10.1007/s00221-008-1611-6

Bernstein, L. E., Auer, Jr., E. T., & Moore, J. K. (2004). Audiovisual speech binding: Convergence or association? In G. Calvert, C. Spence, & B. E. Stein (Eds.), *The handbook of multisensory processes* (pp. 203–223). Cambridge, MA: MIT Press.

Bethge, M., Rotermund, D., & Pawelzik, K. (2003). Optimal neural rate coding leads to multimodal firing rate distributions. *Computational Neural Systems, 14,* 303–319.

Bichot, N. P., & Desimone, R. (2006). Finding a face in the crowd: Parallel and serial neural mechanisms of visual selection. *Programming Brain Research, 155,* 147–156. In C. I. Moore & R. Cao (2008). The hemo-neural hypothesis: On the role of blood flow in information processing. *Journal of Neurophysiology, 99*, 2035–2047.

Bizley, J. K., & King, A. J. (2012). What can multisensory processing tell us about the functional organization of auditory cortex?. In M. M. Murray & M. T. Wallace (Eds.), *The neural bases of multisensory processes* (pp. 31–48). Boca Raton, FL: CRC Press.

Blythe, S. (2003). Meeting the paradox of computer-mediated communication in writing instruction. In P. Takayoshi & B. Huot (Eds.), *Teaching writing with computers: An introduction* (pp. 118–127). Boston, MA: Houghton Mifflin.

Borton, S., & Huot, B. (2007). Chapter 8: Responding and assessing. In C. L. Selfe (Ed.), *Multimodal composition: Resources for teachers* (pp. 99–111). Cresskill, NJ: Hampton Press.

Bouwman, H., van den Hoof, B., van de Wijngaert, L., & Dijk, J. (2005). *Information and communication technology in organizations. Adaptation, implementation, use and effects*. London: Sage.

Boynton, L. (2002). When the class bell stops ringing: The achievements and challenges of teaching online first year composition. *Teaching English in the Two-Year College, 29,* 298–311.

Brady, L. (2001). Fault lines in the terrain of distance education. *Computers and Composition, 18,* 347–358.

Brandt, D. (1999). *Literacy, opportunity, and economic change*. Albany, NY: National Research Center on English Learning & Achievement.

Brandt, D. (2001). *Literacy in American lives*. Cambridge, MA: Cambridge University Press.

Brandt, D. (2005). Writing for a living: Knowledge and the knowledge economy. *Written Communication, 22,* 166–197.

Bremner, A. J., & Spence, C. (2008). Unimodal experience constrains while multi-sensory experiences enrich cognitive construction. *Behavioral and Brain Sciences, 31,* 335–336.

Brooks, R. A., & Stein, L. A. (1994). Building brains for bodies. *Autonomous Robots, 1,* 7–25.

Burgess, N. (2008). Spatial cognition and the brain. *Annals of the New York Academy of Sciences, 1124*, 77–97.

Calvert, G., Spence, C., & Stein, B. (Eds.). (2004). *The handbook of multisensory processes*. Cambridge, MA: MIT Press.

Campos, J. L., & Bulthoff, H. H. (2012). Multimodal integration during self-motion in virtual reality. In M. Murray & M. Wallace (Eds.), *The neural bases of multisensory processes* (pp. 603–628). Boca Raton, FL: CRC Press.

Cao, R. (2011). *The hemo-neural hypothesis: Effects of vasodilation on astrocytes in mammalian neocortex.* Thesis. MIT.

Cargile-Cook, K., & Grant-Davie, K. (Eds.). (2005). *Online education: Global questions, local answers.* Amityville, NY: Baywood.

Carroll, J. M., Smith-Kerker, P. L., Ford, J. R., & Mazur, S. (1988). The minimal manual. In E. Doheny-Farina (Ed.), *Effective documentation: What we have learned from research* (pp. 73–102). Cambridge, MA: MIT Press.

Charney, D., Reder, L., & Wells, G. (1988). Studies of elaboration in instructional texts. In E. Doheny-Farina (Ed.), *Effective documentation: What we have learned from research* (pp. 47–72). Cambridge, MA: MIT Press.

Clark, C. (2013). *Comparison chart—PowerPoint and Prezi. NspireD^2.* Retrieved May 23, 2013 from http://ltlatnd.wordpress.com/2011/03/22/comparison-chart-powerpoint-and-prezi/

Clark, R. E., & Feldon, D. F. (2005). Five common but questionable principles of multimedia learning. In R. E. Mayer (Ed.), *The Cambridge handbook of multimedia learning* (pp. 97–111). Cambridge, MA: Cambridge University Press.

Clemo, H. R., Keniston, L. P., & Meredith, M. A. (2012). Structural basis of multisensory processing: Convergence. In M. Murray & M. Wallace (Eds.), *The neural bases of multisensory processes* (pp. 3–14). Boca Raton, FL: CRC Press.

Cohen, D., & Halgren, E. (2004). Magnetoencephalography. In G. Adelman & B. Smith (Eds.), *Encyclopedia of neuroscience* (1st, 2nd, 3rd eds.). New York, NY: Elsevier.

Colavita, F. B. (1974). Human sensory dominance. *Perception & Psychophysics, 16,* 409–412.

College Composition and Communicative Conference Committee on Best Practices in Online Writing Instruction. (2011). *Report of the Committee for Best Practices in Online Writing Instruction.* Retrieved April 18, 2011 from http://www.ncte.org/library/NCTEFiles/Groups/CCCC/Committees/OWI_State-of-Art_Report_April_ 2011.pdf

Crano, W. D., & Brewer, M. B. (2002). *Principles and methods of social research* (2nd ed.). Mahwah, NJ: Lawrence Erlbaum Associates.

Csibra, G. (2004). Mirror neurons and action observation: Is simulation involved? *Interdisciplines* http://www.interdisciplines.org/mirror/papers/4. In Gallese, V., Eagle, M. N., & Migone, P. (2007) Intentional attunement: Mirror neurons and the neural underpinnings of interpersonal relations. *Journal of the American Psychoanalytic Association, 55,* 131-176.

DeAngelis, G. C., & Angelaki, D. E. (2012). Visual-vestibular integration for self-motion perception. In M. Murray & M. Wallace (Eds.), *The neural bases of multisensory processes* (pp. 629–649). Boca Raton, FL: CRC Press.

Decker, C. L., & Adamek, M. E. (2004). Meeting the challenges of social work research in long term care. *Social Work in Health Care, 38,* 47–65.

DePew, K., Fishman, T., Fahey Ruetenik, B., & Romberger, J. (2006). Designing efficiencies: The parallel narratives of distance education and composition studies. *Computers and Composition, 23,* 49–67.

Dickey, M. D. (2005). Brave new (interactive) worlds: A review of the design affordances and constraints of two 3D virtual worlds as interactive learning environments. *Interactive Learning Environments, 13,* 121–137.

Dinero, D. A. (2005). *Training within industry: The foundation of lean.* New York, NY: Productivity Press.

Donahue, S. E., Woldorff, M. G., & Mitroff, S. R. (2010). Video game players show more precise multisensory temporal processing abilities. *Attention, Perception and Psychophysics, 72,* 1120–1129.

Eastern Washington University. (2008). *Shooting better video composition.* Retrieved November 20, 2009 from http://www.ewu.edu/x3498.xml

Eeles, E., & Rockwood, K. (2008). Delirium in the long term care setting: Clinical and research challenges. *Journal of American Medical Directors Association, 9,* 157–161.

Elmer, M. (2004). Multisensory integration: How visual experience shapes spatial perception. *Current Biology, 14,* R115–R117.

Gaillet, L. L. (2012). (Per)Forming archival research methodologies. *College Composition and Communication, 64,* 35–58.

Gallese, V., Eagle, M. N., & Migone, P. (2007). Intentional attunement: Mirror neurons and the neural underpinnings of interpersonal relations. *Journal of the American Psychoanalytic Association, 55,* 131–176.

Gee, J. P. (1996). *Social linguistics and literacies: Ideology in discourses* (2nd ed.). London, UK: RoutledgeFalmer.

Gee, J. P. (2003). *What video games have to teach us about learning and literacy.* New York, NY: Palgrave Macmillan.

Gee, J. P., Hull, G., & Lankshear, C. (1996). *The new work order: Behind the language of the new capitalism.* Boulder, CO: Westview Press.

Gibbons, A. (2012). *Multimodality, cognition, and experimental literature.* London: Routledge.

Gopnik, A., Meltzoff, A., & Kuhl, P. (1999). *The scientist in the crib: What early learning tells us about the mind.* New York, NY: HarperCollins.

Graupp, P., & Wrona, R. J. (2006). *The TWI workbook: Essential skills for supervisors.* New York, NY: Productivity Press.

Gross C. G. (1998). *Brain vision memory: Tales in the history of neuroscience.* Cambridge, MA: MIT Press.

Gruber, D. (2012). *Neurorhetoric and the dynamism of the neurosciences: Mapping translations of mirror neurons across the disciplines.* Unpublished doctoral dissertation. North Carolina State University, Raleigh, NC. Retrieved from http://www.lib.ncsu.edu/resolver/1840.16/7623

Gutfreund, Y., & Knudsen, E. J. (2004). Visual instruction of the auditory space map in the midbrain. In G. Calvert, C. Spence, & B. E. Stein (Eds.), *The handbook of multisensory processes* (pp. 613–624). Cambridge, MA: MIT Press.

Heath, S. B. (2007, 1983). *Ways with words: Language, life and work in communities and classrooms.* Cambridge, MA: Cambridge University Press.

Heim, M. (1998). *Virtual realism.* New York, NY: Oxford University Press.

Henson, R. N., Flandin, G., Friston, K. J., & Mattout, J. (2010). A parametric empirical Bayesian framework for fMRI-constrained MEG/EEG source reconstruction. In *Human Brain Mapping* (pp. 1512–1531). doi: 10.1002/hbm.20956

Herrington, A., Hodgson, K., & Moran, C. (2009). Technology, change and assessment: What we have learned. In A. Herrington, K. Hodgson, & C. Moran (Eds.), *Teaching the new writing: Technology, change and assessment in the 21st century classroom* (pp. 198–208). New York, NY: Teachers College Press.

Hewett, B. (2004). Generating new theory for online writing instruction. *Kairos*. Retrieved from http://kairos.technorhetoric.net/6.2/features/hewett/

Hewett, B. L., Remley, D., Zemliansky, P., & Dipardo, A. (2010). Frameworks for talking about collaborative writing. In B. L. Hewett & C. Robidoux (Eds.), *Virtual collaborative writing in the workplace. Technologies and processes* (pp. 28–51). Hershey, CT: IGI Global.

Hicks, W. (1973). *Words and pictures.* New York, NY: Arno Press.

Hoiland, E. (2012). *Neuroscience for kids*. Retrieved May 14, 2012 from http://faculty.washington.edu/chudler/plast.html

Holsti, O. R. (1969). *Content analysis for the social sciences and humanities*. Reading, MA: Addison-Wesley.

Howard, I. P., & Templeton, W. B. (1966). *Human spatial orientation*. London, UK: Wiley.

Hudson, K. (2010). Applied training in virtual environments. In W. Ritke-Jones (Ed.), *Virtual environments for corporate education: Employee learning and solutions* (pp. 110–122). Hershey, PA: IGI Global.

Hutchins, E. (1995). *Cognition in the wild.* Cambridge, MA: MIT Press.

Hutchins, E. (2000). Distributed cognition. IESBS Distributed Cognition. Retrieved June 22, 2012 from http://www.artmap-research.com/wp-Content/uploads/2009/11/Hutchins_DistributedCognition.pdf. May 18, 2000.

Huynh, H. K., Beers, C., Willemsen, A., Lont, E., Laan, E., Dierckx, R., et al. (2012). High-intensity erotic visual stimuli de-activate the primary visual cortex in women. *Journal of Sexual Medicine*. doi: 10.1111/j.1743-6109.2012.02706.x

Jack, J. (2012). *Neurorhetorics*. New York, NY: Psychology Press.

Jack, J., & Appelbaum, L. G. (2010). "This is your brain on rhetoric:" Research directions for neurorhetorics. *Rhetoric Society Quarterly, 40*, 411–437.

Jacob, P., & Jeannerod, M. (2004). The motor theory of social cognition: A critique. *Interdisciplines* http://www.interdisciplines.org/mirror/papers/2. In Gallese, V., Eagle, M. N., & Migone, P. (2007). Intentional attunement: Mirror neurons and the neural underpinnings of interpersonal relations. *Journal of the American Psychoanalytic Association, 55*, 131-176.

Jaschik, S. (2010). When writing class moves online. *Inside Higher Ed*. Retrieved March 18, 2011 from http://www.insidehighered.com/news/2010/03/19/writing. March 19, 2010.

Johnson, T. (2008). How to create video tutorials. I'dratherbewriting.com. Retrieved November 20, 2009 from http://www.idratherbewriting.com/2008/09/11/how-i-create-video-tutorials/

Jones, J. (2007). The necessity of teaching video composition. *Viz.: Visual rhetoric, visual culture, pedagogy*. University of Texas. Retrieved November 20, 2009 from http://viz.cwrl.utexas.edu/taxonomy/term/34

Kajikawa, Y., Falchier, A., Musacchia, G., Lakatos, P., & Schroeder, C. (2012). Audio-visual integration in nonhuman primates: A window into the anatomy and physiology of cognition. In M. Murray & M. Wallace (Eds.), *The neural bases of multisensory processes* (pp. 65–98). Boca Raton, FL: CRC Press.

Kalning, K. (2007). *If Second Life isn't a game, what is it?* MSNBC.com. Retrieved January 20, 2011 from http:www.msnbc.msn.com/id/17538999/ns/technology_and_science-games/3/12/2007

Kalyuga, S. (2005). Prior knowledge principle in multimedia learning. In R. E. Mayer (Ed.), *The Cambridge handbook of multimedia learning* (pp. 325–336). Cambridge, MA: Cambridge University Press.

Kayser, C., Petkov, C. I., Remedios, R., & Logothetis, N. K. (2012). Multisensory influences on auditory processing: Perspectives from fMRI and electrophysiology. In M. Murray & M. Wallace (Eds.), *The neural bases of multisensory processes* (pp. 99–114). Boca Raton, FL: CRC Press.

Keetels, M., & Vroomen, J. (2012). Perception of synchrony between the senses. In M. Murray & M. Wallace (Eds.), *The neural bases of multisensory processes* (pp. 147–177). Boca Raton, FL: CRC Press.

Khayat, P. S., Spekreijse, H., & Roelfsema, P. R. (2006). Attention lights up new object representations before the old ones fade away. *Journal of Neuroscience, 26,* 138–142. In C. I. Moore & R. Cao (2008). The hemo-neural hypothesis: On the role of blood flow in information processing. *Journal of Neurophysiology, 99*, 2035–2047.

Khoe, W., Freeman, E., Woldorff, M. G., & Mangun, G. R. (2006). Interactions between attention and perceptual grouping in human visual cortex. *Brain Research, 1078,* 101–111. In C. I. Moore & R. Cao (2008). The hemo-neural hypothesis: On the role of blood flow in information processing. *Journal of Neurophysiology, 99*, 2035–2047.

King, A. J., & Calvert, G. (2001). Multisensory integration: Perceptual grouping by eye and ear. *Current Biology, 11*, R322–R325.

King, A. J., Doubell, T. P., & Skaliora, I. (2004). Epigenetic factors that align visual and auditory maps in the ferret midbrain. In G. Calvert, C. Spence, & B. E. Stein (Eds.), *The handbook of multisensory processes* (pp. 599–612). Cambridge, MA: MIT Press.

Kirkevold, M., & Bergland, A. (2007). The quality of qualitative data: Issues to consider when interviewing participants who have difficulties providing detailed accounts of their experiences. *International Journal of Qualitative Studies on Health and Well-Being, 2,* 68–75.

Kress, G. (2003). *Literacy in the new media age.* London, UK: Routledge.

Kress, G. (2009). *Multimodality: A social semiotic approach to contemporary communication.* London, UK: Routledge.

Kress, G., & Van Leeuwen, T. (2006). *Reading images: The grammar of visual design.* London, UK: Routledge.

Kress, G., & Van Leeuwen, T. (2001). *Multimodal discourse: The modes and media of contemporary communication.* London, UK: Arnold.

Krippendorff, K. (2004). *Content analysis: An introduction to its methodology* (2nd ed.). Thousand Oaks, CA: Sage.

Krippendorff, K. (2008). Reliability issues in content analysis data: What it is and why. In K. Krippendorff & M. A. Bok (Eds.), *The content analysis reader* (pp. 350–357). Thousand Oaks, CA: Sage.

Krishna, A., Elder, R. S., & Caldara, C. (2010). Feminine to smell but masculine to touch? Multisensory congruence and its effect on the aesthetic experience. *Journal of Consumer Psychology, 20,* 410–418.

Lacey, S., & Sathian, K. (2012). Representation of object form in vision and touch. In M. Murray & M. Wallace (Eds.), *The neural bases of multisensory processes* (pp. 179–187). Boca Raton, FL: CRC Press.

Lemke, J. L. (1998). Multiplying meaning: Visual and verbal semiotics in scientific text. In J. R. Martin & R. Veel (Eds.), *Reading science: Critical and functional perspective on discourses of science* (pp. 87–114). London, UK: Routledge.

Lemke, J. L. (1999). Discourse and organizational dynamics: Website communication and institutional change. *Discourse and Society, 10,* 21–47.

Lewkowicz, D. J., & Kraebel, K. S. (2004). The value of multisensory redundancy in the development of intersensory perception. In G. Calvert, C. Spence, & B. E. Stein (Eds.), *The handbook of multisensory processes* (pp. 655–678). Cambridge, MA: MIT Press.

Li, J., D'Souza, D., & Yunfei, D. (2011). Exploring the contribution of virtual worlds to learning in organizations. *Human Resource Development Review, 10,* 264–285.

Liker, J., & Meier, D. (2007). *Toyota talent: Developing your people the Toyota way.* New York, NY: McGraw-Hill.

Linden Labs. (2010, September 5). *Machinima content companies.* Retrieved January 21, 2011 from http://wiki.secondlife.com/wiki/Machinima_Content_Companies.

Lowood, H. (2005). Real-time performance: Machinima and game studies. *The Digital Media and Arts Association Journal, 2,* 10–17. Retrieved January 20, 2011 from http://www.idmaa.org/journal/pdf/iDMAa_Journal_Vol_2_No_1_screen.pdf

MacNealy, M. S. (1999). *Strategies for empirical research in writing.* Boston, MA: Allyn and Bacon.

Malaby, T. M. (2007). Contriving constraints: The gameness of Second Life and the persistence of scarcity. *Innovations: Technology, Governance, Globalization, 2*(3), 62–67.

Markel, M. (2010). *Technical communication* (9th ed.). Boston, MA: Bedford/St. Martin's.

Massumi, B. (2002). *Parables for the virtual.* Durham, NC: Duke University Press.

Matthiessen, C. M. I. M. (2007). The multimodal page: A system functional exploration. In T. D. Royce & W. L. Bowcher (Eds.), *New directions in the analysis of multimodal discourse* (pp. 1–62). Mahwah, NJ: Lawrence Erlbaum Associates.

Mayer, R. E. (2001). *Multi-media learning.* Cambridge, MA: Cambridge University Press.

Mayer, R. E. (Ed.). (2005). *The Cambridge handbook of multimedia learning.* Cambridge, MA: Cambridge University Press.

McCraty, R., Atkinson, M., Tomasino, B. A., & Bradley, R. T. (2009). The coherent heart heart–brain interactions, psychophysiological coherence, and the emergence of system-wide order. *Integral Review, 5,* 2. Retrieved March 3, 2014 from http://integral-review.org/documents/McCraty%20et%20al,,%20Coherent%20Heart,%20Vol.%205%20No.%202.pdf

Mesko, B. (2008). Unique medical simulation in Second Life. *Science Roll.* Retrieved December 30, 2011 from http://scienceroll.com/2008/08/17/unique-medical-simulation-in-second-life/. August 17, 2008.

Mitchell, W. J. T. (1995). *Picture theory*. Chicago, IL: University of Chicago Press.

Mollman, S. (2007). Wii+Second Life=New training simulator. *Wired*. Retrieved December 31, 2011 from http://www.wired.com/gadgets/miscellaneous/news/2007/07/wiimote. July 27, 2007.

Moore, C. I., & Cao, R. (2008). The hemo-neural hypothesis: On the role of blood flow in information processing. *Journal of Neurophysiology, 99*(5), 2035–2047.

Morain, M., & Swarts, J. (2012). YouTutorial: A framework for assessing instructional online video. *Technical Communication Quarterly, 21,* 6–24.

Moreno, R., & Mayer, R. E. (2000). A learner-centered approach to multimedia explanations: Deriving instructional design principles from cognitive theory. *Interactive Multimedia Electronic Journal of Computer-Enhanced Learning, 2.*

Munhall, K. G., & Vatikiotis-Bateson, E. (2004). Spatial and temporal constraints on audiovisual speech perception. In G. Calvert, C. Spence, & B. E. Stein (Eds.), *The handbook of multisensory processes* (pp. 177–188). Cambridge, MA: MIT Press.

Murray, E. A., Sheets, H. A., & Williams, N. A. (2010). The new work of assessment: Evaluating multimodal compositions. *Computers and Composition Online*. Retrieved March 23, 2010 from http://www.bgsu.edu/cconline/murray_etal/index.html

Murray, J. (2009). *Non-discursive rhetoric: Image and affect in multimodal composition*. New York, NY: SUNY Press.

Murray, M. M., & Wallace, M. T. (Eds.). (2012). *The neural bases of multisensory processes*. Boca Raton, FL: CRC Press.

Neal, M. (2010). Assessment in the service of learning. *College Composition and Communication, 61,* 746–758.

Neal, M. (2011). *Writing assessment and the revolution in digital texts and technologies*. New York, NY: Teachers College Press.

Nelson, D. (1995). *Farm and factory: Workers in the midwest 1880-1990*. Bloomington, IN: Indiana University Press.

New London Group. (1996). A pedagogy of multiliteracies: Designing social futures. *Harvard Educational Review, 66*, 60–92.

Newell, F. N. (2004). Cross-modal object recognition. In G. Calvert, C. Spence, & B. E. Stein (Eds.), *The handbook of multisensory processes* (pp. 123–139). Cambridge, MA: MIT Press.

Norman, D. A. (1988). *The design of everyday things*. New York, NY: Doubleday.

Norman, D. (1999). Affordance, conventions and design. *Interactions*. Retrieved December 2, 2009 from http://www.jnd.org/dn.mss/affordance_conv.html

Norman, D. (2004). Affordance, conventions and design. Retrieved December 2, 2009 from http://www.jnd.org/dn.mss/affordance_conv.html

Odell, L., & Katz, S. M. (2009). "Yes, a t-shirt!": Assessing visual composition in the writing class. *College Composition and Communication, 61,* W197–W216.

Odell, L., & Katz, S. (Eds.). (2012). Assessing multimedia. *Technical Communication Quarterly* (special issue), *21.*

Oliu, W. E., Brusaw, C. T., & Alred, G. J. (2010). *Writing that works* (10th ed.). Boston, MA: Bedford/St. Martin's.

Pasqualotto, A., & Proulx, M. J. (2012). The role of visual experience for the neural basis of spatial cognition. *Neuroscience and Biobehavioral Reviews, 36,* 1179–1187.

Perelman, C., & Olbrechts-Tyteca, L. (1969). *The new rhetoric: A treatise on argumentation* (J. Wilkinson & P. Weaver, Trans.). Notre Dame, IN: University of Notre Dame Press.

Perrault, T. J., Rowland, B. A., & Stein, B. E. (2012). The organization and plasticity of multisensory integration on the midbrain. In M. Murray & M. Wallace (Eds.) *The neural bases of multisensory processes* (pp. 279–300). Boca Raton, FL: CRC Press.

Petroski, H. (1996). *Invention by design: How engineers get from thought to thing.* Cambridge, MA: Harvard University Press.

Pillay, S. S. (2012). *Your brain and business. The neuroscience of great leaders.* Upper Saddle River, NJ: Pearson/Financial Press.

Pinker, S. (1997). *How the mind works*. New York, NY: W. W. Norton and Sons.

Ramsay, I. S. et al. (2013). Affective and executive network processing associated with persuasive antidrug messages. *Journal of Cognitive Neuroscience, 25*(7), 1136–1147. doi: 10.1162/jocn_a_00391. Epub 2013 Mar 26.

Reid, A. (2007). *The two virtuals: New media and composition*. West Lafayette, IN: Parlor Press.

Remley, D. (2009) Training within industry as short-sighted community literacy-appropriate training program: A case study of a worker-centered training program. *Community Literacy Journal, 3*(2), 93–114.

Remley, D. (2010). Developing digital literacies in Second Life: Bringing Second Life to business writing pedagogy and corporate training. In W. Ritke-Jones (Ed.), *Virtual environments for corporate education: Employee learning and solutions* (pp. 169–193). Hershey, PA: IGI Global.

Remley, D. (2011). The practice of assessing multimodal PowerPoint slide shows. *Computers and Composition Online.* http://www.bgsu.edu/cconline/CCpptassess/index.html

Remley, D. (2012). Forming assessment of machinima video. *Computers and Composition Online.*

Remley, D. (2013). Templated pedagogy: Factors affecting standardized writing pedagogy with online course management systems. *Journal of Writing and Pedagogy 5*(1), 105–120.

Remley, D. (2014). *Exploding technical communication: Workplace literacy hierarchies and their implications for literacy sponsorship*. Amityville, NY: Baywood.

Rice, R. E. (1993). Media appropriateness: Using social presence theory to compare traditional and new organizational media. *Human Communication Research, 19*, 451–484.

Richards, A. R. (2003). Argument and authority in visual representations of science. *Technical Communication Quarterly, 12,* 183–206.

Ritke-Jones, W. (2010). *Virtual environments for corporate education: Employee learning and solutions*. Hershey, PA: IGI Global.

Rivers, N. A. (2011). Future convergences: Technical communication research as cognitive science. *Technical Communication Quarterly, 20,* 384–411.

Rizzolatti, G., Fadiga, L., Fogassi, L., & Gallese, V. (1996). Premotor cortex and the recognition of motor actions. *Cognitive Brain Research, 3,* 131–141.

Rodabaugh, J. H. (1995). Ohio and World War II. In T. Smith (Ed.), *An Ohio reader: Reconstruction to the present* (pp. 314–318). Grand Rapids, MI: William B. Eerdman Publishing.

Rowley-Joliet, E. (2004). Different visions, different visuals: A social semiotic analysis of field-specific visual composition in scientific conference presentations. *Visual Communications, 3*(2), 145–177.

Royce, T. D. (2007). Intersemiotic complimentary: A framework for multimodal discourse analysis. In T. D. Royce & W. L. Bowcher (Eds.), *New directions in the analysis of multimodal discourse* (pp. 63–109). Mahwah, NJ: Lawrence Erlbaum Associates.

Sathian, K., Prather, S. C., & Zhang, M. (2004). Visual cortical involvement in normal tactile perception. In G. Calvert, C. Spence, & B. E. Stein (Eds.), *The handbook of multisensory processes* (pp. 703–709). Cambridge, MA: MIT Press.

Schiappa, E. (1993). *Defining reality: Definitions and the politics of meaning.* Carbondale, IL: Southern Illinois University Press.

Schnotz, W. (2005). An integrated model of text and picture comprehension. In R. E. Mayer (Ed.), *The Cambridge handbook of multimedia learning* (pp. 49–60). Cambridge, MA: Cambridge University Press.

Scribner, S., & Cole, M. (1981). *The psychology of literacy.* Cambridge, MA: Harvard University Press.

Selfe, C. (2004). Toward new media text: Taking up the challenges of visual literacy. In A. F. Wysocki, J. Johnson-Eilola, C. Selfe, & G. Sirc (Eds.), *Writing new media: Theory and applications for expanding the teaching of composition* (pp. 67–110). Logan, UT: Utah State Press.

Shams, L., Kamitani, Y., & Shimojo, S. (2004). Modulations of visual perception by sound. In G. Calvert, C. Spence, & B. E. Stein (Eds.), *The handbook of multisensory processes* (pp. 27–33). Cambridge, MA: MIT Press.

Shrestha, L. B. (2006). Life expectancy in the United States. *CRS Report for Congress.* Washington, DC: Congressional Research Service.

Silverman, D. (2006). *Interpreting qualitative data* (3rd ed.). London, UK: Sage.

Sorapure, M. (2005). Between modes: Assessing student new media compositions. *Kairos, 10*(2).

Soto-Franco, S., & Vljamae, A. (2012). Multisensory interactions during motion perception: From basic principles to media applications. In M. Murray & M. Wallace (Eds.), *The neural bases of multisensory processes* (pp. 583–602). Boca Raton, FL: CRC Press.

Spence, C., Parise, C., & Chen, Y. (2012). The Colavita visual dominance effect. In M. Murray & M. Wallace (Eds.), *The neural bases of multisensory processes* (pp. 529–556). Boca Raton, FL: CRC Press.

Stern, L. A., & Solomon, A. (2006). Effective faculty feedback: The road less traveled. *Assessing Writing, 11,* 22–41.

Stratton, R. C. (1943). Explosion while handling fuzed twenty-pound fragmentation bomb clusters (Correspondence report). August 13, 1943.

Strauss, A. L. (1987). *Qualitative analysis for social scientists.* Cambridge, MA: Cambridge University Press.

Street, B. (1984). *Literacy in theory and practice.* Cambridge, MA: Cambridge University Press.

Sullivan, P. (2001). Practicing safe visual rhetoric on the World Wide Web. *Computers and Composition, 18,* 103–122.

Taub, H. (1980). Informed consent, memory and age. *The Gerontologist, 20,* 686–690.

Teston, C. (2012). Moving from artifact to action: A grounded investigation of visual displays of evidence during medical deliberations. *Technical Communication Quarterly, 21,* 187–209.

Tufte, E. (1990). *Envisioning information.* Cheshire, CT: Graphics Press.

Tufte, E. (2003). *The cognitive style of PowerPoint.* Cheshire, CT: Graphics Press.

Tufte E. R. (1983). *The visual display of quantitative information.* Cheshire, CT: Graphics Press.

Tufte E. R. (2006). *Beautiful evidence.* Cheshire, CT: Graphics Press.

Turing A. M. (1950). Computing machinery and intelligence. *Mind Quarterly Review Psychological Philosophy, 59,* 433–460.

TWI Learning Partnership. (2015). http://www.twilearningpartnership.com/TWI jobmetrics.asp

University of Wisconsin-Stout. (2010). *PowerPoint rubric.* Retrieved September 3, 2010 from http://www.uwstout.edu/static/profdev/rubrics/pptrubric.html

Unsworth, L. (2007). Multiliteracies and multimodal text analysis in classroom work with children's literature. In T. D. Royce & W. L. Bowcher (Eds.), *New directions in the analysis of multimodal discourse* (pp. 331–359). Mahwah, NJ: Lawrence Erlbaum Associates.

van der Wall, E. E., & van Gilst, W. H. (2012). Neurocardiology: Close interaction between heart and brain. *Netherland Heart Journal, 21*(2), 51–52. Retrieved March 3, 2014 from http://www.ncbi.nlm.nih.gov/pmc/articles/PMC3547430/

Van Horn, J. et al. (2012, May 16). Mapping connectivity damage in the case of Phineas Gage. *PLoS One* doi: 10.1371/journalpone.0037454

van Leeuwen, T. (2003). A multimodal perspective on composition. In T. Ensink & C. Sauer (Eds.), *Framing and perspectivising in discourse.* Amsterdam: Benjamins.

Vie, S. (2008). Tech writing, meet *Tomb Raider*: Video and computer games in the technical communication classroom. *E-Learning, 5*(2. Retrieved January 19, 2011 from www.wwwords.co.uk/ELEA

Volkow, N. D. et al. (2007). Depressed dopamine activity in caudate and preliminary evidence of limbic involvement in adults with attention-deficit/hyperactivity disorder. *General Psychology, 64*(8), 932–940. doi: 10.1001/archpsyc.64.8.932

Vygotsky, L. (1978). *Mind in society: The development of higher psychological processes.* M. Cole, S. Scribner, V. John-Steiner, & E. Souberman (Eds.). Cambridge, MA: Harvard University Press.

Wagner, M. (2007, April 26). Using Second Life as a business-to-business tool. *InformationWeek: The Business Value of Technology.* Retrieved September 10, 2010 from http://www.informationweek.com/blog/main/archives/2007/04/using_second_li_2.html;jsessionid=KIHDIBAKMQZBDQE1GHPSKH4ATMY32JVN

Walbert, D. (n.d.). Evaluating multimedia presentations. Learn NC. Retrieved September 3, 2010 from http://www.learnnc.org/lp/pages/647

Wallace, M. T. (2004). The development of multisensory integration. In G. Calvert, C. Spence, & B. E. Stein (Eds.), *The handbook of multisensory processes* (pp. 625–642). Cambridge, MA: MIT Press.

Wallace, M. T., Perrault, J. R. T. T., Hairston, D., & Stein, B. (2004). Visual experience is necessary for the development of multisensory integration. *The Journal of Neuroscience, 27,* 9580–9584.

War Manpower Commission. (1945). *Training within Industry Report 1940-1945.* Washington, DC: U.S. Government.

Welch, R., & Warren, D. (1986). Intersensory interactions. In K. Boff, L. Kauffman, & J. Thomas (Eds.), *Handbook of perception and human performance: Vol I. Sensory processes and human performance.* New York, NY: Wiley.

Whithaus, C. (2012). Claim-evidence structures in environmental science writing: Modifying Toulmin's model to account for multimodal arguments. *Technical Communication Quarterly, 21*, 105–128.

Wolbers, T., Hegarty, M., Büchel, C., & Loomis, J. (2008). How the brain keeps track of changing object locations during observer motion. *Nature Neuroscience, 11,* 1223–1230.

Worden, M. S., Foxe J. J., Wang, N., & Simpson, G. V. (2000). Anticipatory biasing of visuospatial attention indexed by retinotopically specific alpha-band electroencephalography increases over occipital cortex. *Journal of Neuroscience, 20,* RC63. In C. I. Moore & R. Cao (2008). The hemo-neural hypothesis: On the role of blood flow in information processing. *Journal of Neurophysiology, 99*, 2035–2047.

Wysocki, A. F. (2001). Impossibly distinct: On form/content and word/image in two piece computer-based interactive media. *Computers and Composition, 18*, 207-234.

Wysocki, A. F. (2004). Opening new media to writing: Openings and justifications. In A. F. Wysocki, J. Johnson-Eilola, C. L. Selfe, & G. Sirc (Eds.), *Writing new media: Theory and application for expanding the teaching of composition* (pp. 1–42). Logan, UT: Utah State University Press.

Yin, R. K. (1984). *Case study research: Design and methods.* Newbury Park, CA: Sage.

Index

Page references followed by f indicate a figure.